The Genius Checklist

The Genius Checklist

Nine Paradoxical Tips on How You Can Become a Creative Genius!

Dean Keith Simonton

The MIT Press
Cambridge, Massachusetts
London, England

This book was set in Stone Serif by Westchester Publishing Services. Printed and bound in the United States of America.

Library of Congress Cataloging-in-Publication Data
Names: Simonton, Dean Keith, author.
Title: The genius checklist : nine paradoxical tips on how you can become a creative genius / Dean Keith Simonton.
Description: Cambridge, MA : MIT Press, [2018] | Includes bibliographical references and index.
Identifiers: LCCN 2017050207 | ISBN 9780262038119 (hardcover : alk. paper)
Subjects: LCSH: Genius. | Creative ability.
Classification: LCC BF412 .S553 2018 | DDC 153.9/8--dc23
LC record available at https://lccn.loc.gov/2017050207

10 9 8 7 6 5 4 3 2 1

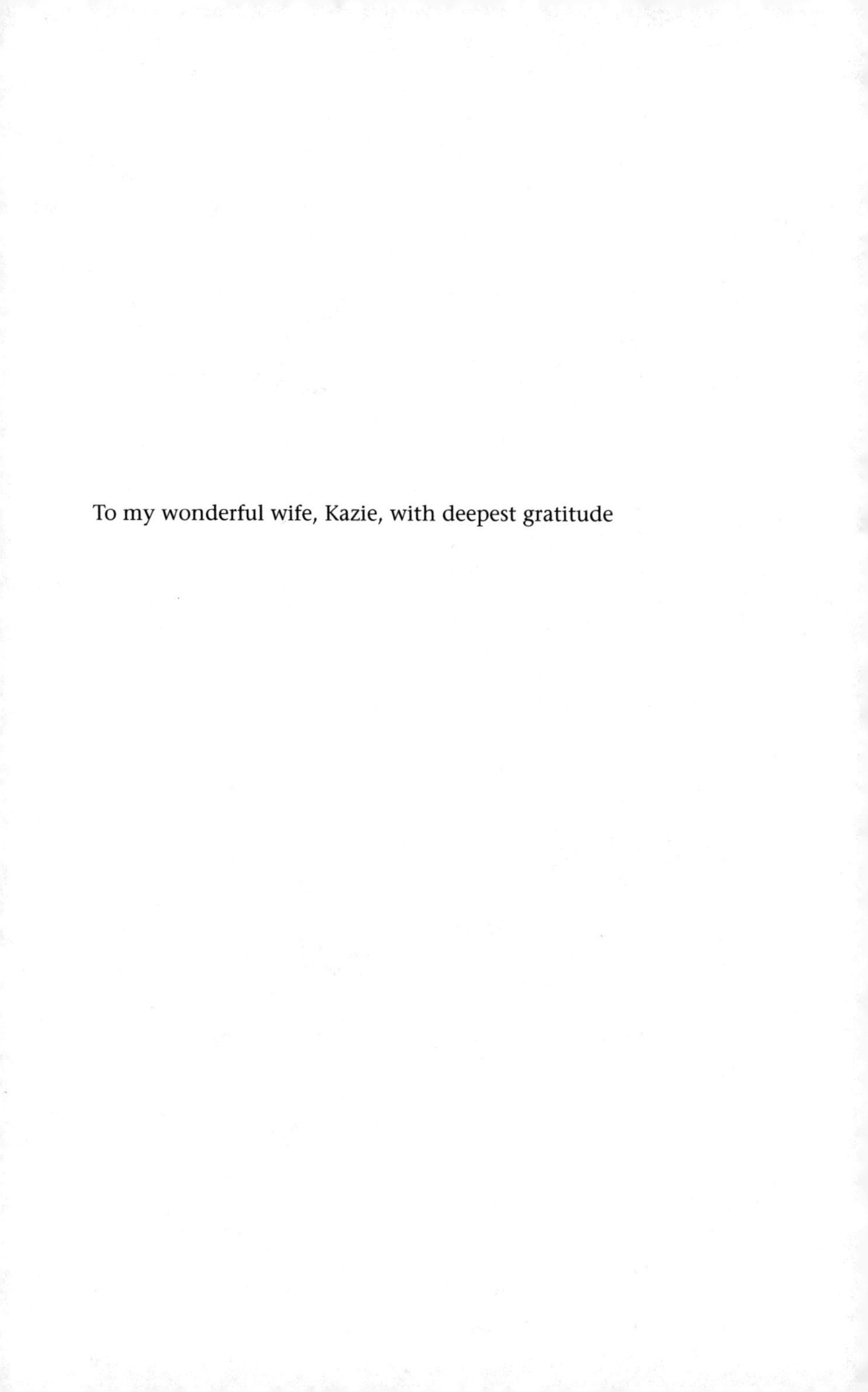

To my wonderful wife, Kazie, with deepest gratitude

Contents

Prologue

Not long ago I published a rather speculative essay in *Nature* about whether genius had become obsolete in the current natural sciences. Without my consent, the journal's editors changed my original title to another that was much more provocative: "After Einstein: Scientific Genius Is Extinct." I seemed to be claiming that geniuses had ceased to exist! My email inbox was soon inundated with reactions both supportive and critical. The supportive group encompassed a diversity of opinions, such as those who argued for the "end of science" as a creative enterprise altogether. Yet the critical group was much more intriguing because it most often involved irate protests from self-proclaimed "neglected geniuses." A typical example was someone who claimed to have completely overturned Einstein's physics, but who was still impatiently waiting for a call from Stockholm to schedule the Nobel Prize ceremony. If Einstein was universally acclaimed as a genius, this respondent justified, why wouldn't someone receive the same acclamation for proving Einstein wrong? After all, $E=mc^2$ should actually be $E=mQ^2$; just don't make the mistake of asking what Q stands for!

Sometimes these overlooked geniuses expected me to certify their claims. Wasn't I a presumed expert on the science of genius? When I declined to validate their genius status with some impressive-looking certificate or at least a rubber stamp, a few accused me of being an absolute fraud and went so far as to post their accusations on their personal websites (where their earthshaking work is most often published). One made it very clear that he expected me to be fired by my university as soon as the hoax I perpetuated had been revealed. Being an expert on genius thus seems to have a downside: too many out there are seeking vindication and expecting it now! None of them want to wait around for posthumous fame.

You don't have to be a crackpot to place so much value on genius. In 1981 the MacArthur Foundation began its Fellows Program to honor highly creative people with a substantial amount of money with no strings attached (presently $625,000 paid over five years). Journalists immediately proclaimed the Fellowship a "Genius Grant," thus certifying the recipients as genuine geniuses. Both brick and internet bookstores contain self-help books with titles like *Discover Your Genius, Awakening Your Inner Genius*, and *The Secret Principles of Genius*. Just follow their proffered pearls of wisdom and your own genius is guaranteed. Of course, many parents hope that one of their children is a "budding genius," sometimes even trying to accelerate the budding by exposing their infant to some "Baby Einstein" product. Genius is so highly valued in contemporary society that the honor is often loosely applied to achievements that don't necessarily require superlative intellectual capacity—such as outstanding athletes. What does it mean to apply the designation "basketball genius" to a point guard like Stephen Curry? Sure, he has broken almost all records for field goals beyond the arc, but does that really signify something more than supreme talent?

Although the word *genius* is often bandied about in popular culture and mass media, it is most frequently used in complete ignorance of the extensive scientific research on that very subject. These investigations began more than 150 years ago and continue to the present day. Of course, these research findings are not accessible to general readers. Most results are buried in technical journals full of esoteric statistics and mathematics. So somehow the central discoveries must be extracted and communicated to a broader audience. That is one major goal of this book. The principal way of achieving this end is to make ample use of concrete illustrations and anecdotes drawn from the lives of well-known creative geniuses: true stories instead of numbers and equations—stories that actually illustrate what we now know. With that in mind, I have also adopted an informal method of citation in these pages—keyed to phrases rather than note numbers. It emphasizes and documents original research on the subject of (and related to) creative genius rather than the historical and biographical information easily researched by a Google search.

Yet another goal is just as important: the need to convey the complexity of the scientific results, but without making the presentation too complicated. Genius operates in ways sometimes so subtle as to seem contradictory. For example, is genius born or made? The correct scientific answer is both born *and* made. But too often authors will choose sides, arguing for example that genius is entirely made rather than born. To avoid such oversimplification, I have translated the scientific results into nine paradoxes, thus representing both sides rather than just one. Contradictory though the resulting tips may seem, both arguments contain a grain of truth. Indeed, knowing when one aspect or the other holds is crucial to understanding the intricacies of genius, such as why the "correctness" of the tip often depends on the contrast between scientific and artistic genius. What kind of genius do you want to be? A reincarnated Albert Einstein or the latest Pablo Picasso? The answer makes a big difference in determining the most appropriate advice.

The net outcome is a list of nine paradoxical tips that constitute a sort of double-edged checklist about how to assess whether you or anybody else counts as a genius. Hence, read carefully!

Tip 1
Score at Least 140 on an IQ Test / Don't Even Bother Taking the Test!

Everybody knows that geniuses have high IQs, right? So just take a regular intelligence test and see what you get. Seems very straightforward. Given all of the purported IQ tests available on the internet—often for free—anyone should at least be able to obtain an approximate score. But what specific number do you need? What's the cutoff between counting as a genuine genius and just being intellectually bright?

One way to answer this question is do a Google search. If you put in "genius IQ," for example, lots of interesting results pop up on the screen, including one site titled "20 Celebrities With Genius IQs." After the assertion that a genius has an IQ of 135 or higher, the following "stars" from a variety of fields are said to qualify: Matt Damon 135, Jodie Foster 138, Natalie Portman 140, Shakira 140, Madonna 140, Nicole Kidman 142, Steve Martin 143, Arnold Schwarzenegger 144, David Duchovny 147, Ben Stein 150, Lisa Kudrow 154, Sharon Stone 154, Dr. Mehmet Oz 158, Ashton Kutcher 160, Quentin Tarantino 160, Conan O'Brien 160, Mayim Bialik 163, Kris Kristofferson 166, Dolph Lundgren 166, and James Woods 184. That last score is quite impressive. According to the website, fewer than 100 people in the entire United States would score as high as Wood does—a super genius! Odds are that everybody he meets is much less intelligent. Do they feel proportionately intimidated?

Unfortunately, no information is given about the chosen IQ threshold, or even what tests the celebrities took. Certainly all 20 would qualify for membership in Mensa, an international society that requires an IQ in the top 2% of the population. What this minimum means exactly depends on the specific IQ test, but this elite percent would often assume an IQ of 130–132 on the most common measures.

Yet others might argue that Mensa is a bit too generous to guarantee genius status. Bona fide genius must exhibit a much higher IQ. Of course, this argument for increased selectivity most likely comes from somebody who scored much higher than 135. (Kutcher, Tarantino, and O'Brien, for instance, might even set the qualification at IQ 160.) Anyway, in support of raising the standards, the *American Heritage Dictionary* defines a genius as a "person who has an exceptionally high intelligence quotient, typically above 140." Although the dictionary doesn't specify the IQ test behind this particular number, on many tests this score would place a person in the top 1% of the population. Poor Damon and Foster would not make the cut. But where does this particular threshold come from in the first place? It turns out that the answer to that question involves quite a story—one lasting half a century.

Terman's Stanford-Binet and His 1,500+ "Termites"

People too often forget that IQ tests haven't been around that long. Indeed, such psychological measures are only about a century old. Early versions appeared in France with the work of Alfred Binet and Theodore Simon in 1905. However, these tests didn't become associated with genius until the measure moved from the Sorbonne in Paris to Stanford University in Northern California. There Professor Lewis M. Terman had it translated from French into English, and then standardized on sufficient numbers of children, to create what became known as the Stanford-Binet Intelligence Scale. That happened in 1916. The original motive behind these tests was to get a diagnostic to select children at the lower ends of the intelligence scale who might need special education to keep up with the school curriculum. But then Terman got a brilliant idea: Why not study a large sample of children who score at the top end of the scale? Better yet, why not keep track of these children as they pass into adolescence and adulthood? Would these intellectually gifted children grow up to become genius adults?

Must Adult Geniuses Start Life as High-IQ Boys and Girls?

Terman subjected hundreds of school kids to his newfangled IQ test. This is where the IQ 140 cutoff came into play. Obviously, he didn't want a sample so large that it would be impractical to follow their intellectual development. Taking the top 2% of the population would clearly yield a group twice as large as the top 1%. Moreover, a less select group might be less prone to become geniuses. So why not catch the crème de la crème?

The result was a group of 1,528 extremely bright boys and girls who averaged around 11 years old. And to say they were "bright" is a very big understatement. Their average IQ was 151, with 77 claiming IQs between 177 and 200. These children were subjected to all sorts of additional tests and measures, repeatedly so, until they reached middle age. The result was the monumental *Genetic Studies of Genius*, five volumes appearing between 1925 and 1959, although Terman died before the last volume came out. These highly intelligent people are still being studied today, or at least the small number still alive. They have also become affectionately known as "Termites"—a clear contraction of "Termanites."

Now comes the bad news: none of them grew up to become what many people would consider unambiguous exemplars of genius. Their extraordinary intelligence was channeled into somewhat more ordinary endeavors as professors, doctors, lawyers, scientists, engineers, and other professionals. Two Termites actually became distinguished professors at Stanford University, eventually taking over the longitudinal study that included themselves as participants. Their names are Robert R. Sears and Lee Cronbach—and nowhere are they as well-known as Ivan Pavlov, Sigmund Freud, or Jean Piaget, three obvious geniuses in the history of psychology.

Furthermore, many Termites failed to become highly successful in any intellectual capacity. These comparative failures were far less likely to graduate from college or to attain professional or graduate degrees, and far more likely to enter occupations that required no higher education whatsoever. We're talking only of the males here, too. It would be unfair to consider the females who were born at a time in which all women were expected to become homemakers, no matter how bright. (Even among those women with IQs exceeding 180, not all pursued careers.) Strikingly, the IQs of the successful men did not substantially differ from the IQs of the unsuccessful men. Whatever their differences, intelligence was not a determining factor in those who made it and those who didn't.

What Are the Chances for Sub-genius Boys and Girls?

The story goes from bad to worse. Of the many rejects—the children with tested IQs not high enough to make it into the Terman sample—at least two attained higher levels of acclaim than those who had the "test smarts" to become Termites. Here are their stories:

Luis Walter (Luie) Alvarez was born in San Francisco, just up the peninsula from Stanford. He was around 10 years old when he took Terman's test but scored too low to enter the sample. Yet that rejection did not prevent him from getting his PhD at age 25 from the University of Chicago. Even as a graduate student he began to make important contributions to physics, eventually becoming "one of the most brilliant and productive experimental physicists of the twentieth century." One manifestation of this brilliance was his work on hydrogen bubble chambers for studying elementary particles, which led to his receiving the 1968 Nobel Prize in Physics. No Termite received the Nobel, in physics or otherwise. Oops! As if to embarrass his supposed intellectual superiors even more, Luie also worked on several wartime projects, including involvement in the Manhattan Project that produced the first plutonium bomb ("Fat Man") and work on radar systems for landing aircraft under adverse conditions—for which he won a trophy from the National Aeronautic Association. Going well beyond his core discipline, he even tackled such problems as how to determine the number of shots fired during the assassination of John F. Kennedy and how to use the muons generated by cosmic rays to detect hidden chambers in the Egyptian pyramids. Most remarkably, he collaborated with his geologist son Walter Alvarez to advance the "Alvarez hypothesis," purporting that the mass extinction of the dinosaurs was caused by a massive asteroid impact about 66 million years ago. It's probably safe to say that anybody who has a serious fascination with the sudden demise of the dinosaurs is thoroughly familiar with this family conjecture.

William (Bill) Shockley is the second Termite reject who went on to attain the Nobel Prize in Physics, which he shared with two colleagues in 1956. Born just one year before Alvarez, he grew up in Palo Alto near Stanford, the university his mother had graduated from. Despite his sub-genius score on Terman's IQ test, he managed to get his BS from Cal Tech and his PhD from MIT, both prestigious technical institutions. He then joined Bell Labs and began to publish extensively in solid-state physics, getting his first patent at age 28. Like Luie, Bill got involved in the World War II effort, especially with respect to radar (in his case, bomb sights). His contributions earned him the war secretary's Medal for Merit in the same year that Alvarez got his trophy. Although Shockley did not participate in the Manhattan Project, he was asked to estimate the casualties that would most likely result were the Allies to invade the Japanese mainland. His horrific estimates (for both sides) helped sway the decision to drop atomic bombs on Hiroshima and Nagasaki. After the war, he returned to Bell Labs, where the goal was to find a solid-state substitute for the old glass vacuum tubes that then dominated electronics. The upshot was the transistor. In the same year that he received his Nobel, Shockley moved to Mountain View, California, not far from the Stanford campus. Coincidentally, his move received encouragement from Stanford's dean of engineering, who was Terman's son! There he founded the Shockley Semiconductor Laboratory in what later became known as Silicon Valley. (Terman died in the same year, perhaps unaware of the phenomenal achievements of this second IQ reject.) Sadly, Shockley was an ineffective business manager, and like Steve Jobs much later at Apple, got pushed out of his own company, but permanently so. Yet he was immediately hired as a distinguished professor of engineering and applied science at his mother's alma mater, where he ended his career.

So there we have it: quite in line with the second half of Tip 1, little Luie and Bill could have skipped taking the Stanford-Binet and still claim achievements that surpassed Terman's IQ-certified "geniuses." But they are not unique among Nobel laureates. Both James Watson, the co-discoverer of DNA's structure, and Richard Feynman, who worked on the path integral of quantum mechanics, had scores too low to gain membership in Mensa. So who needs to take an IQ test?

Testing the IQ of Dead People: Cox's 301 Geniuses

At the start of this chapter I mentioned doing a Google search to drag up the IQs of 20 super-bright celebrities. All were still alive at the time of the search, and presumably all had actually taken some version of a recognized IQ test. But often such searches yield results that are seemingly impossible: IQ scores for geniuses long, long deceased before even the earliest tests were ever invented. For instance, I found the following list on one website, with the geniuses in ascending order of IQ: Austrian composer Wolfgang Amadeus Mozart 165; British scientist Charles Darwin 165; German philosopher Immanuel Kant 175; French mathematician René Descartes 180; Italian scientist Galileo Galilei 185; French mathematician Blaise Pascal 195; British political scientist John Stuart Mill 200; and German poet and writer Johann Wolfgang von Goethe 210. The scores are very impressive. The overlap of these IQ results with those of the celebrities given earlier is small. Creative genius reigns over celebrated talent. Yet how are these second set of scores even possible? Every single one of the listed geniuses died long before 1905, when the most preliminary intelligence test first emerged!

Amazingly, these IQs for the long deceased are intimately connected with the IQs of the Termites. The connection comes from the following true story.

Some years after Terman had begun his study of 1,528 high-IQ boys and girls, he acquired a new graduate student named Catharine Cox. Because her mentor's investigation was already well in progress, she found it difficult to carve out a portion that might serve as her doctoral dissertation. So she tried a bold alternative. If Terman was going to see if high-IQ kids grew up to become adult geniuses, why not do the opposite? In particular, why not pick a group of obvious adult geniuses, and then try to assess their childhood and adolescent IQs retrospectively from their biographies?

Coming up with a list of geniuses is the easy part. For example, nowadays we would just google "famous scientists" or "famous artists" (try it). In Cox's pre-internet era, the equivalent would be to compile a list from biographical dictionaries and other (paper) reference works. Fortunately, she found an already published list, from which she extracted the most famous names. She ended up with 301 historic creators and leaders (192 and 109, respectively). No doubt that her sample included some of the top figures in the history of modern Western civilization. Besides the eight mentioned above, she would study big-time creators like Isaac Newton, Jean-Jacques Rousseau, Miguel de Cervantes, Ludwig van Beethoven, and Michelangelo (as well as leaders like Napoleon Bonaparte, Horatio Nelson, Abraham Lincoln, and Martin Luther)—all of them folks who can boast extensive Wikipedia biographies.

The hard part was estimating IQ scores for all 301 geniuses. How is that even possible?

Happily, Terman had already shown, just one year after devising the Stanford-Binet test, how IQs could be estimated from biographies. Back in those days IQ was defined as a literal "intelligence quotient," namely a child's mental age divided by his or her chronological age, the arithmetic result then multiplied by 100. The mental age is determined by performance on intellectual tasks that are age graded. Accordingly, if a 5-year-old could do well on tasks more suitable for 10-year-olds, the IQ quotient would become 200 (= 10/5 × 100). Pretty straightforward, no?

Terman applied this method to the early intellectual development of one of his heroes, Francis Galton, the very first scientist to investigate genius. For instance, Francis wrote the following little letter to his older sister: "I am 4 years old and I can read any English book. I can say all the Latin Substantives and Adjectives and active verbs besides 52 lines of Latin poetry. I can cast up any sum in addition and can multiply by 2, 3, 4, 5, 6, 7, 8, [9], 10, [11]. I can also say the pence table. I read French a little and I know the clock." The two numbers in brackets had been obscured, one by an erasure that made a hole and the other by a more effective paper patch. The young Galton apparently saw that he was claiming too much—an act that itself could be considered evidence of a higher mental age. Now what is the normal expectation for 4-year-old children? Only this: be able to give their gender; name a key, knife, and penny placed before them; repeat back three numbers just told them, and compare two lines in front of their eyes. That's it! Galton, were he average, would not even be able to count four coins until age 5, give his age until 6, copy a written sentence until 7, or write from dictation until 8. In any event, using additional biographical evidence like this, Terman inferred that Galton's IQ approached 200. His mental age was almost twice his chronological age.

Cox decided to apply the same method to the 301 but, moving beyond her mentor's scope, added methodological improvements such as compiling detailed chronologies of intellectual growth from multiple biographical sources and having independent raters make IQ estimates from those chronologies. The resulting doctoral dissertation was very impressive: not only did she earn her Stanford PhD but Terman also decided to have it incorporated as the second volume of *Genetic Studies of Genius*. It became the only volume out of the eventual five that did not involve the Termites, and the only volume that Terman did not author or co-author. This study's inclusion is important because Terman had only published the first volume the year before, in 1925, and the remaining volumes were long in the future, the last one not appearing until 1959. The Termites had to grow up after all: the 10-year-olds had to become middle-aged.

Now comes the big question: Does the 301-genius study by Cox endorse the supreme importance of a high IQ? Or does her study argue for its relative irrelevance? In brief, do you really need to take the test? Let's see!

Pro: High IQ Is Essential to Fame and Fortune!

As a group, no doubt, the 301 boasted genius-level IQs, scores that clearly excelled those received by the Termites. Their average IQ ranged between 153 and 164, depending on the specific estimation adopted (e.g., ages 0–16 versus ages 17–26). What makes Cox's case especially powerful is that she doesn't just present her IQ estimates, but abstracts the raw biographical data on which those estimates were based. Readers can thereby make up their own mind. For instance, like Galton, Mill's mental age was about twice his chronological age. Don't believe it? Then ponder the following facts about his early education:

> John Stuart Mill had no childhood; his interests and his activities were mature from the first. ... He began to learn Greek at 3; and from then until his 9th year studied Greek classics, making daily reports of his reading. ... At 7 he read Plato; at 8 he began the study of Latin. Before the end of the year he was busily reading the classical Latin writers. He did not neglect mathematics: at 8 his course included geometry and algebra; at 9 conic sections, spherics, and Newton's arithmetic were added. ... At 10 and 11 both mathematical and classical studies were continued; astronomy and mechanical philosophy were also included. In fluxions [calculus], begun at 11, Mill was largely self-taught.

Besides his extremely accelerated formal education, Mill would engage in behaviors indicative of a rather precocious mind. For example, he wrote a history of Rome at age 6½. At what age do most people write their first history of Rome—or anything else for that matter?

Yet Cox took an additional step. Not all of her creative geniuses achieved the same magnitude of eminence. On the contrary, many were also-rans who would most likely be unknown except to cognoscenti. Examples include the French philosopher Antoine Arnauld, the Swedish chemist Jöns Jacob Berzelius, and the Scottish writer William Robertson. At the same time, her geniuses sometimes exhibited sub-genius IQs—at times too low even to qualify for Mensa. Among these less stratospheric intellects are creators like the Spanish writer Miguel de Cervantes, the Polish astronomer Nicolaus Copernicus, and the French painter Nicolas Poussin. Because all of her geniuses had already been previously ranked on achieved eminence according to the amount of space devoted to them in the reference works—the French general Napoleon came out # 1 while the English writer Harriet Martineau ranked # 301 (ouch!)—Cox could easily correlate the IQ scores with the ranks (inverted, of course). She obtained a statistically significant correlation, and the correlation remained significant even after correcting for data reliability (meaning that the biographical information wasn't equally good for all geniuses). Furthermore, this positive relation has been replicated multiple times since her own 1926 demonstration. Hence, achieved eminence is associated with superlative intelligence. Her mentor Terman thus seems vindicated!

Con: High IQ Is Tangential to Fame and Fortune!

So far, so good. The first half of Tip 1 seems justified: scoring high in IQ would seem to increase the odds of attaining acclaim. That being said, four problems cast some doubt on this conclusion.

Problem #1: The Intelligence-Eminence Correlation The relation between IQ and achieved eminence is not huge or even large. Most statisticians would classify it as a "moderate" relationship. In practical terms, that means that there's ample room for exceptions at either end. The highly eminent can have IQs lower than average and supremely high IQs can be associated with relative obscurity. I've already given three examples of the former, so who illustrates the latter? How about Paolo Sarpi, the Venetian historian? Although his estimated IQ was as high as 195, making him one of the very brightest among the 301, his eminence ranking put him in the lower 20%, that is, 242nd!

A more contemporary example is Marilyn vos Savant, who was once listed in the *Guinness Book of World Records* as having the highest recorded IQ. Reportedly, she had taken a revised version of the Stanford-Binet when she was just 10 years old, and got a perfect score! Although there's some debate about how best to translate that performance into a precise IQ estimate, it is certainly arguable that she is more intelligent than the brightest Termite and any member of Cox's 301. Yet what is her main accomplishment? Becoming famous for her super-high IQ! Exploiting that distinctive status, she writes the Sunday column "Ask Marilyn" for *Parade* magazine. That column doesn't come close to the writing in *Don Quixote* or *On the Revolutions of the Celestial Spheres*, which her two intellectual inferiors, Cervantes and Copernicus, managed to pull off! An extra 60 IQ points or more didn't buy her any creative edge at all.

Problem #2: The Creative Domain IQ's relevance to achievement appears dependent on the domain of achievement. Some domains seem to place far less emphasis on intelligence relative to other domains. For example, famous leaders tend, on the average, to have lower IQs than famous creators. The low IQs of commanders (generals and admirals) is really conspicuous—in Cox's 301, about 20 points lower than everybody else! The most distinguished military leader in the sample was certainly Napoleon, yet with the highest estimate for his IQ at only 145 he would have been among the less intelligent Termites. Sometimes an excessively high IQ can work against effective leadership: too much of a good thing. Being a "man (or woman) of the people" often implies having an intellect closer to their level. Comprehension is more persuasive than competence. No wonder, then, that the presidents of the United States don't do much better than the commanders! These results not only help us understand why the epithet *genius* seems more likely to be assigned to great creators than to great leaders, but also provide a justification for largely ignoring the latter group in this book. Leaders may exhibit charisma, perhaps, but creators are more likely to display genius.

Problem #3: Personality and Persistence Matter Because the IQ-eminence correlation is so low, even if positive, other psychological factors must be involved that have nothing to do with intelligence. Cox herself revealed as much. Besides assessing her 301 geniuses on IQ, she also took a subset of 100 geniuses for whom the biographical data was particularly good and then measured them on 67 personality traits. Motivational traits emerged as especially critical—persistence standing out above the rest. As she put it: "High but not the highest intelligence, combined with the greatest degree of persistence, will achieve greater eminence than the highest degree of intelligence with somewhat less persistence." In a sense, the highly eminent are overachievers, attaining more distinction that would be expected from their IQs alone. Oddly, this result echoes what Galton had argued more than a half century earlier:

> By natural ability, I mean those qualities of intellect and disposition, which urge and qualify a man to perform acts that lead to reputation. I do not mean capacity without zeal, nor zeal without capacity, nor even a combination of both of them, without an adequate power of doing very laborious work. But I mean a nature which, when left to itself, will, urged by an inherent stimulus, climb the path that leads to eminence, and has strength to reach the summit—one which hindered or thwarted, will fret and strive until the hindrance is overcome, and it is again free to follow its labour-loving instinct.

Natural ability entails not just intelligence, but both passion and perseverance—or what some contemporary psychologists call "grit."

Problem # 4: Deceptive Assessment Cox cheated! Not deliberately, I mean, but she cheated nonetheless. Her IQ scores cannot really be equated with Terman's IQ scores. It's not just that one set is too high or too low relative to the other, but that the two sets don't really measure the same thing, at least not most of the time. On the one hand, the Stanford-Binet gauges a person's acquisition and development of basic cognitive skills, such as memory and reasoning, and rudimentary scholastic skills, such as the proverbial three Rs of "reading, 'riting, and 'rithmetic." Almost everybody would be expected to possess those elementary skills by the time they reach adulthood. What makes one person smarter than another is mainly the speed at which those skills are acquired. A 5-year-old with an IQ of 200 has somehow managed to master what the average person wouldn't get until age 10, but otherwise there's little difference. On the other hand, Cox's IQ estimates were very often based on skills that would be very rare even in grownups. Because these skills are highly specific to a particular domain of creativity, the resulting scores would contrast like apples and oranges, or perhaps even celery and onions.

To illustrate, consider Mozart. As noted earlier, his IQ is supposedly as high as 165. But what is that estimate based on? Primarily on his musical development as a keyboardist and composer. In music, he was phenomenally precocious. For example, Mozart first started to write little pieces around age 5, and published his first works at age 7. "Between the ages of 7 and 15 he composed works for pianoforte and violin, pianoforte concertos, masses, and church music, 18 symphonies, 2 operettas, and at age 14, an opera." Moreover, by age 6 he also began his notable musical tours throughout Western Europe. One of these concert trips landed Mozart in London, where the child prodigy attracted so much attention that he became the subject of a scientific study later published in the *Philosophical Transactions of the Royal Society*. Mozart's precocious musical skills were not mere hearsay! Yet we have to ask: What does it mean to calculate his "mental age" from these musical achievements? Does it even make sense to specify the typical age when a person composes an opera or goes on a solo concert tour? Obviously not. Those are accomplishments that the vast majority of persons—even most musicians—never manage to realize in their entire lifetimes.

Worse still, outside of music, Mozart's personal development was not nearly so advanced. The issue was raised in the *Transactions* article about whether Mozart's father had deliberately exaggerated his son's youth as a circus-like marketing ploy. It just didn't seem plausible to observers that someone so young could prove so astonishingly proficient. The prodigy could even outdo his father on the imposed tests. Yet not only was the birthdate confirmed as a matter of public record, but the researcher noted that Mozart looked and acted his chronological age: "whilst he was playing to me, a favourite cat came in, upon which he immediately left his harpsichord, nor could we bring him back for considerable time." In addition, little Mozart "would also sometimes run about the room with a stick between his legs by way of horse." Given that 8-year-olds often play on stick horses, and that children start riding them from age 3 up, Mozart's IQ might have been estimated as about 100—*if* his musical talents were completely ignored. Outside of music, and unlike Mill mentioned earlier, Mozart definitely experienced a childhood.

Imagine how well you might do on an IQ test if you only answered the questions on which you do best? That's why Cox was indirectly but inadvertently cheating!

Tested Intelligence or Achieved Eminence? It's Your Choice!

In most dictionaries, the entry for "genius" provides multiple definitions. And getting a score of 140 on an IQ test is not the only one. Here's another also given by the *American Heritage Dictionary*: "Native intellectual power of an exalted type, such as is attributed to those who are esteemed greatest in any department of art, speculation, or practice; instinctive and extraordinary capacity for imaginative creation, original thought, invention, or discovery." That definition definitely fits the top creators in Cox's 301, but it fails to apply to any of Terman's 1,500+ Termites. Hence, the first tip in our genius checklist has this paradoxical advice. If you're smart enough to score 140 or better on an IQ test, then all by all means go that route. Given that you can take this test when you're only 2 years old, this may be the best choice if doable. A 2-year-old doing what 3-year-olds can do is not that difficult. You can get it over with while still a toddler and then spend the rest of your life basking in the glory of certified geniushood.

But if you don't succeed, even after multiple retesting, there's no need to despair. Just pick some "department of art, speculation, or practice," and then achieve eminence for some "imaginative creation, original thought, invention, or discovery." Admittedly, this second course seems much more arduous, and may even take a whole lifetime to accomplish, but at least you can avoid taking any IQ test whatsoever! Plus, your claim to genius status just might withstand the test of time. Authentic genius leaves an impact longer than a testing session, creating a pervasive impression that endures for decades, even centuries.

Tip 2

Go Stark Raving Mad / Become the Poster Child for Sanity!

Everybody fascinated with genius has heard many tragic stories about the ones succumbing to recurrent madness. The Dutch painter Vincent van Gogh suffers from diverse psychopathological symptoms for much of his life—once infamously cutting off part of one ear—and finally shoots himself in the chest (as we've long been told, although recent scholarship suggested foul play), only to die 30 hours later. English writer Virginia Woolf endures frequent bouts of debilitating depression until she writes her husband a suicide note, loads her overcoat pockets with stones, and walks into a nearby river, her drowned body not found until more than two weeks later. The repeated depressive episodes of the American poet Sylvia Plath lead her to numerous suicide attempts—such as overdosing on drugs and driving her car into a river—before she puts her head in an oven and dies from carbon monoxide poisoning. The latter becomes the poison of choice of fellow American poet Anne Sexton, albeit by running her car engine in a closed garage. The tragedies go on and on. Such stories provide dramatic support for the popular image of the "mad genius."

To be sure, the relation between suicide and mental illness is complex. On the one hand, persons may kill themselves for causes that have nothing to do with mental or emotional disorder. Indeed, in some cultures, from ancient Rome to medieval Japan, suicide offered a rational means to an honorable death. Petronius, the Roman author of the pornographic *Satyricon*, when accused of treason, avoided execution by cutting open his veins and slowly bleeding to death—all the while pleasantly conversing and dining with friends! At other times suicide provides an escape from a progressive disease and thus might be better conceived as a form of self-euthanasia. The American comedian and actor Robin Williams experienced drug and alcohol problems most of his life, but not until he faced the intensifying adversity of Lewy body dementia did he decide to hang himself. Sometimes it's better to end it all before any free choice becomes impossible.

On the other hand, mental illness does not have to end in suicide. Sometimes creative geniuses endure their on-again off-again symptoms for their entire lives and then die unexpectedly of conditions unrelated to their mental health. A well-known example is depicted in the 2001 film *A Beautiful Mind*, which deals with the paranoid schizophrenia that plagued the American mathematician John Forbes Nash Jr. The Nobel laureate and his wife died in a car crash while taking a taxi home from the airport after traveling to Norway to receive the prestigious Abel Prize for Mathematics. Other instances include many of those creative geniuses who experienced alcoholism, drug abuse, or both—addictions contributing to their cause of death. One famous alcoholic, the French artist Henri de Toulouse-Lautrec, often devised ingenious ways to ensure that he always had a drink literally at hand—most notably by hollowing out his cane and filling it with liquor. This genius's alcoholism, combined with syphilis, eventually killed him at age 36, but he was not a suicide.

Notwithstanding the apparent ease with which we can identify unquestionably suicidal, alcoholic, depressed, and schizophrenic geniuses, such anecdotes cannot possibly demonstrate that genius is necessarily linked to madness. At best, such specific cases merely prove that mental illness need not prevent someone from becoming a creative genius. So don't count yourself out of the running just because you undergo psychopathological episodes of one kind or another. Perfect mental health is *not* a prerequisite for the job. Just ask Van Gogh, Woolf, Nash, or Toulouse-Lautrec. Even so, is it possible that extremely *imperfect* mental health might actually prove an asset for an aspiring creative genius? This question raises the nasty "mad-genius controversy" that has been raging for centuries. Where some psychologists insist on an essential link, others argue that the very concept of "mad genius" represents a pure myth if not outright hoax.

Sadly, debates in psychology too often adopt either/or positions. Yet both sides can be right—but right in different ways. And that's the case here. What the antagonists seldom realize is that the question "Is genius connected to madness?" more precisely encompasses rather separate questions. Moreover, the answer to one question does not necessarily constrain the answers to the other questions. The issues are logically independent of each other, or "orthogonal" in formal terms. Three such orthogonal questions are perhaps the most critical:

First, how does the mental health of creative geniuses compare with the mental health of those who can't qualify for that honor? That is, how does the risk for psychopathology differ between the genius-grade creator and a person neither creative nor a genius?

Second, how much does the mental health of creative geniuses depend on the domain in which the geniuses make their contributions? To offer a specific example, are artistic geniuses more inclined toward cognitive or emotional disorder than are scientific geniuses?

Third, how does the risk of psychopathology change as a function of the degree of creative achievement? Are the supremely eminent more prone to symptoms than their far lesser known but still creative colleagues?

Again, the answers to these three questions are totally unrelated to each other. For example, creative geniuses might be less prone to mental illness, but artistic geniuses might still be more susceptible to illness than scientific geniuses. Similarly, the greatest creative geniuses may be more vulnerable to psychopathology even when lesser creative geniuses have a higher likelihood of exceptional mental health. The mutual independence of these questions will become more evident in the three sections that follow.

Creative Genius versus the Uncreative Non-Genius

Most researchers who debate the mad-genius issue probably do so with the first of those three questions in mind: *How does the mental health of the creative genius stack up against the mental health of the uncreative non-genius?* Creative geniuses differ from the rest of us in part because they are more likely to suffer from mental illness. Some are even "stark raving mad," as suggested by the first part of the Tip 2 paradox. Geniuses like the German composer Robert Schumann even end their lives in a "mental institution" or "asylum for the insane." Yet, obviously, had Schumann become institutionalized at the beginning of his career and remained so until the end, it is most unlikely that his surname would appear in this paragraph. He would have had no creative achievements to put on his résumé. So is timing everything? Make sure you don't go mad until you finish at least a masterpiece or two? If so, the English poet Thomas Chatterton barely made it. Already revealing suicidal thoughts, he managed to attract some attention with his (forged) "Rowley poems," and then committed suicide by arsenic poisoning at age 17. He thereby became an iconic hero of the Romantic era—the tragic, misunderstood artist who dies miserably alone in his attic study!

To provide a suitable scientific response to this first question, I'll start by discussing two key issues that must be resolved at the outset, and then turn to two characteristic empirical studies that provide further insight.

Two Key Issues Concerning Creative Genius and Mental Illness

First of all, what do we even mean by mental illness or psychopathology? These terms subsume many different types of illness; the most recent edition of the *Diagnostic and Statistical Manual of Mental Disorders* (*DSM-5*), which was published by the American Psychiatric Association in 2013, lists well over a dozen, such as the schizophrenic, bipolar, depressive, anxiety, obsessive-compulsive, dissociative, and personality disorders. Moreover, these disorders are defined by a multitude of symptoms that can vary in both frequency and intensity. How often do you feel depressed? When you do succumb to depression, does it involve no more than "feeling the blues" that make you mope around the house all day, or do you become profoundly suicidal, going so far as to write a suicide note?

As the last example implies, many symptoms can operate along a continuum at subclinical levels, indicating a personal propensity for a disorder without the actual incapacitating effects of that disorder. Most of us can experience mood changes, anxiety, obsessions and compulsions, and perhaps even mild delusions, such as wishful thinking and fanciful overconfidence. No hard and fast line can be drawn between normal and abnormal. Even a discrete sign of illness, like committing suicide, features ambiguities. Maybe the suicide attempt was not really intended to result in death, but merely to communicate a "cry for help." One or more of Plath's attempted suicides may have had the latter motivation—perhaps even, not that we'll ever know, her final, successful attempt. Such pleas, however, are sometimes at the mercy of fate: for example, the bus that brings a loved one home from work every afternoon "like clockwork" might run late on a particular day, with an unintended lethal outcome. If a suicide attempt was not designed to succeed, but does so anyway, does that make the act less pathological? If so, by how much? And who decides?

Second, what research subjects or participants best define the two sides of the comparison between the creative genius and the uncreative non-genius?

On the one side, much psychological research is devoted to the relationship between scores on so-called creativity tests and performance on standard diagnostic measures (such as the Minnesota Multiphasic Personality Inventory; or MMPI) or personality instruments assessing some traits associated with subclinical levels of psychopathology (such as the Eysenck Personality Questionnaire; or EPQ). Genius, per se, is not required. Indeed, the research participants are very often no more than college undergraduates who just so happen to be taking an introductory psychology class. Still, the mad-genius controversy cannot be resolved without incorporating the genius part. Even selecting people who score high on creativity tests who just so happen to have high IQs will not give you a creative genius. If vos Savant received a high score on some creativity test, she would still not count as a "creative genius" because of the stipulation with which we closed Tip 1. She hasn't (yet?) made her mark with some "imaginative creation, original thought, invention, or discovery" in a "department of art, speculation, or practice."

On the other side, with whom are the creative geniuses to be compared? Any random person on the street? Like maybe you or me? Or individuals carefully matched in gender, ethnicity, age, education, socioeconomic class, and any other demographic variable connected to the risk of psychopathology? Given that many creative geniuses lived in very different times and places, how is that even possible? Who would represent the matched controls for Newton, Rousseau, Cervantes, Michelangelo, and Beethoven? In short, establishing an appropriate base rate is by no means easy to do. But the risk level for the comparison group is absolutely crucial to settling the first question. If the expected risk is too low, then the mad-genius hypothesis may receive confirmation, but if too high, then it may receive disconfirmation.

If we take the general population of humanity as the baseline, then we can use the base rate provided by a nationally representative sample that estimated the lifetime prevalence of any mental disorder to be about 50%. No doubt that's a very rough ballpark figure, and perhaps even a slight overestimate, but it's better to err on the conservative side when investigating this controversial question. So how does this rate compare with that for creative genius?

Two Characteristic Studies of Creative Genius and Mental Illness

The following two investigations provide reasonable even if (as always) tentative answers to the core question of the mad-genius debate:

First, Arnold Ludwig, an American psychiatrist, carefully assessed psychopathology in more than a thousand highly eminent creators and leaders (all of them worthy of major biographies). His subjects included manifest creative geniuses such as those listed in table 2.1.

Each and every one satisfies a requirement of the dictionary definition given in Tip 1 (namely, making major contributions to an important domain). Ludwig then used biographical materials to assess the occurrence of various psychopathologies. He broke these assessments down into the different domains of achievement, including a breakdown for the "lifetime rate of any disorder." The latter would seem most compatible with the baseline general population given earlier; if so, almost all of the creative domains exhibit rates above 50%, and many of them quite substantially so. The only apparent exceptions are the creative geniuses in the natural sciences, a point that I will return to in the next section—where it's clearly more relevant.

Second, a British psychiatrist, Felix Post, conducted another investigation on roughly the same question. In his case, he collected biographical data on 291 highly eminent figures, 245 of whom represented scientists, thinkers, writers, artists, and composers rather than politicians, his group of leaders. Although the sample somewhat overlaps Ludwig's, Post's sample is shifted a bit toward the 19th century, and thus includes earlier geniuses like the English scientist Charles Darwin, the German thinker Arthur Schopenhauer, the American writer Herman Melville, the French

Table 2.1
Representative Creative Geniuses in Ludwig's Study of Mental Illness

Poets, novelists, story writers, and playwrights: Guillaume Apollinaire, W. H. Auden, Simone de Beauvoir, Berthold Brecht, Andre Breton, Albert Camus, Truman Capote, Anton Chekhov, Agatha Christie, Jean Cocteau, Joseph Conrad, Noel Coward, E. E. Cummings, Gabrielle D'Annunzio, Arthur Conan Doyle, Theodore Dreiser, T. S. Eliot, William Faulkner, E. M. Forester, Anatole France, Robert Frost, Federico Garcia Lorca, Maxim Gorky, Knut Hamsun, Thomas Hardy, Ernest Hemingway, Hermann Hesse, Alfred Edward Housman, Aldous Huxley, Henrik Ibsen, Henry James, James Joyce, Franz Kafka, Rudyard Kipling, D. H. Lawrence, C. S. Lewis, Sinclair Lewis, Robert Lowell, Maurice Maeterlinck, Andre Malraux, Thomas Mann, Katherine Mansfield, Somerset Maugham, Vladimir Nabokov, Sean O'Casey, Eugene O'Neill, George Orwell, Boris Pasternak, Ezra Pound, Marcel Proust, Rainer Maria Rilke, Carl Sandburg, George Bernard Shaw, Edith Sitwell, John Steinbeck, Johan August Strindberg, Dylan Thomas, Leo Tolstoy, Mark Twain, H. G. Wells, Oscar Wilde, Tennessee Williams, William Carlos Williams, Thomas Wolfe, and William Butler Yeats.

Painters, photographers, sculptors, and architects: Ansel Adams, Diane Arbus, Mary Cassatt, Paul Cézanne, Edgar Degas, Marcel Duchamp, Paul Gaugin, Alberto Giacometti, George Grosz, Edward Hopper, Gustav Klimt, Oskar Kokoshka, Kathe Kollwitz, Le Corbusier, Henri Matisse, Ludwig Mies van der Rohe, Edward Munch, Georgia O'Keeffe, Francis Picabia, Pablo Picasso, Jacob Camille Pissarro, Jackson Pollock, Pierre August Renoir, Diego Rivera, Auguste Rodin, Alfred Stieglitz, Louis Sullivan, Henri Toulouse-Lautrec, Andy Warhol, James Abbott McNeil Whistler, and Frank Lloyd Wright.

Popular and classical composers: George Antheil, Louis Armstrong, Bela Bartok, Alban Berg, Claude Debussy, Antonín Dvořák, Duke Ellington, George Gershwin, Edward Grieg, Paul Hindemith, Leos Janacek, Jerome D. Kern, John Lennon, Gustav Mahler, Charlie Parker, Cole Porter, Sergei Prokofiev, Giacomo Puccini, Sergei Rachmaninoff, Maurice Ravel, Arnold Schoenberg, Alexander Scriabin, Dmitri Shostakovich, Richard Strauss, Igor Stravinsky, Arthur Sullivan, Edgard Varèse, Giuseppe Verdi, Anton von Webern, and Kurt Weill.

Scientists and inventors: Alexander Graham Bell, Niels Bohr, Luther Burbank, George Washington Carver, Marie Curie, Harvey Cushing, Thomas Alva Edison, Albert Einstein, Alexander Fleming, Henry Ford, Robert Goddard, Ernest Everett Just, Charles Kettering, Alfred Charles Kinsey, Ernest Orlando Lawrence, Bill Lear, Joseph Lister, J. Robert Oppenheimer, Albert Szent-Gyorgyi, Nikola Tesla, Alan Turing, Orville Wright, and Wilbur Wright.

Philosophers and theologians: John Dewey, Reinhold Niebuhr, Friedrich Nietzsche, Josiah Royce, Bertrand Russell, George Santayana, Jean-Paul Sartre, Albert Schweitzer, Paul Tillich, and Alfred North Whitehead.

Source: Ludwig 1995 (compiled from appendix A).

artist Gustave Courbet, and the Hungarian composer Franz Liszt. Post also devised a somewhat different approach toward gauging psychopathology. Using an earlier version of the *DSM*, Post actually assessed the magnitude of illness (whatever that specific illness might be). In particular, Post used a rough ordinal scale of *none, mild, marked,* and *severe*. If we just take the last two diagnoses as perhaps most compatible with the baseline comparison, then again creative geniuses exceed the 50% rate, with the scientists once more providing a striking exception. Naturally, if we adopt the top three levels, namely mild, marked, and severe, then the psychopathology separation between creative geniuses and hoi polloi would increase all the more.

Hence, Post's inquiry obtained pretty much the same overall answer as Ludwig's. This congruence is all the more important given that the two studies were apparently carried out in complete ignorance of the each other's work. The investigations were executed almost simultaneously on opposite ends of the Atlantic and originally published in separate scientific journals in both Great Britain and the United States. Plus, neither cites the other's work. Yet because their samples of creative geniuses partly overlap, the two investigations can be considered partial replications of each other.

Some might doubt whether it's possible to conduct posthumous diagnoses using biographical materials. Even so, we already saw in Tip 1 that Catharine Cox was able to extract reliable measures of both IQ and character traits from the same kinds of data sources. In fact, many examples of at-a-distance personality assessment have been developed appreciably since Cox's time, rendering it a well-established technique provided appropriate methodological precautions are observed.

Moreover, a glance at the creative geniuses purported to suffer from mental illness will show that the two psychiatrists' diagnoses could not be far off. For example, Post's list of those assigned to the severe category includes such blatant cases as:

- The American novelist Ernest Hemingway, whose alcoholism, depression, and increasing paranoia only ended when he killed himself with a shotgun. His mental condition may actually have been aggravated by the electroconvulsive therapy that he received at the Mayo Clinic, an intervention that must by itself indicate some disorder.
- The Russian composer and pianist Sergei Rachmaninoff, who experienced depressive episodes that could obstruct creative work. His highly successful Piano Concerto no. 2 was actually dedicated to the psychiatrist who helped him out of one incapacitating occasion of "writer's block."
- The French philosopher Auguste Comte, who entered a mental health hospital for treatment by the famous psychiatrist Jean-Étienne Dominique Esquirol. But Comte left before Esquirol had cured him, and attempted suicide a year later by leaping from a bridge into the Seine.
- The Austrian physicist Ludwig Boltzmann, whose depression worsened to the point that he resigned his academic position. He then went on vacation with his wife and daughter, and hung himself. Not only was that quite a shock for his family, but also it may have denied him full recognition for his Nobel Prize–quality work on statistical mechanics.
- The Norwegian painter Edvard Munch, whose masterpiece *The Scream* provides an iconic image of how it feels to suffer from extreme anxiety, as Munch himself did. He was also afflicted by alcoholism, and he was prone to hallucinations and paranoid

delusions. These symptoms eventually became so cruel that he thought he was going mad, and thus entered a clinic for extensive therapeutic treatment.

It does not even take an MD with a residency in psychiatry to infer that these creative geniuses lived in a psychological world sadly remote from even ordinary mental health. They certainly weren't "flourishing," the term often used in today's Positive Psychology movement! Any attempt to consider creative genius as the epitome of mental health is stymied by cases like these. Creativity is not necessarily even good therapy. Confessional poets who write about their most personal traumas too often end up committing suicide—like Plath and Sexton.

Artistic Genius versus Scientific Genius

Let's take a look at the second question, about how much the mental health of the creative genius depends upon the domain in which he or she makes a mark. As I noted in the previous section, scientific genius does not fit the general pattern seen in the other guises of creative genius. According to Ludwig's data, the lifetime rate of any disorder is only 28% in the natural sciences, markedly below the 50% baseline adopted for comparison. In contrast, the rates are 60% in music composing, 72% in nonfiction, 73% in art, 74% in theater, 77% in fiction, and a whopping 87% in poetry! Of the creative domains, only two have risk rates close to the baseline, namely, architecture at 52% and social sciences at 51%. Post's data yield similar results, even if he divides up the domains differently. Only about 18% of his scientists exhibited severe psychopathology, a figure that contrasts markedly with the 26% for thinkers, 31% for composers, 38% for artists, and 46% for writers. The contrasts work at the other end of the scale as well: 31% of Post's scientists displayed no psychopathology whatsoever, and the percentages decrease accordingly, with 17% for composers, 15% for artists, 10% for thinkers, and merely 2% for writers! That last 2% figure covers only one creative genius, namely Guy de Maupassant, the French short-story writer. Still, even he suffered from mental disorders driven by his later awareness of the deteriorating symptoms of the syphilis that he caught early in life. These disorders led him to attempt suicide by cutting his throat, an act that directly motivated his commitment to the asylum where he died. Because of the organic origin of his death, Post decided to give this creative writer a pass. But would you?

In any event, it seems that the mad-genius idea only really applies to geniuses in the visual and literary arts, and to a lesser extent music composition. Scientific geniuses tend to display more stable mental health than do the others. Even in Post's analysis, 55% of the eminent scientists had either no psychopathology or just mild psychopathology. At the same time, Ludwig's findings suggest that practitioners of at least one art form (architecture) and of one science (social) do not lean toward either mental health or mental illness, but rather fall pretty close to the selected baseline. Neither here nor there.

As always happens in scientific research, the picture gets even more complicated—but also somehow more orderly. There's hidden method in the overt madness.

Paradigmatic Science, Scientific Revolutions, and Psychopathology

Now we can focus a little bit more on a central difference between the natural and social sciences. The two groups of scientific endeavors are not the same. On the one hand, creativity in the natural sciences—such as physics, chemistry, and biology—is most often constrained by paradigms. The term *paradigm* comes from Thomas Kuhn's classic *The Structure of Scientific Revolutions*, where he explains it as follows:

> Some accepted examples of actual scientific practice—examples which include law, theory, application, and instrumentation together—provide models from which spring particular coherent traditions of scientific research. These are the traditions which the historian describes under such rubrics as "Ptolemaic astronomy" (or "Copernican"), "Aristotelian dynamics" (or "Newtonian"), "corpuscular optics" (or "wave optics"), and so on. The study of paradigms, including many that are far more specialized than those named illustratively above, is what mainly prepares the student for membership in the particular scientific community with which he will later practice. ... [Scientists] whose research is based on shared paradigms are committed to the same rules and standards for scientific practice. That commitment and the apparent consensus it produces are prerequisites for normal science, i.e., for the genesis and continuation of a particular research tradition.

On the other hand, the social sciences—such as psychology and sociology—are considered "pre-paradigmatic," and thus creators in those domains operate under fewer constraints. Sometimes, even, "anything goes." Because of this difference in paradigmatic guidance, the risk of mental illness would most likely be higher in the social than the natural sciences, just as Ludwig's data show. Social scientists experience far stronger uncertainties about whether they've even accomplished anything of permanence.

But Kuhn also observed that sometimes paradigms stop working. As more failed predictions or explanations accumulate—the accumulation of what he called "anomalies"—the natural science enters a crisis phase not that different from the permanent state of the social sciences at their best. Eventually, it is hoped, a revolutionary scientist emerges who provides a new paradigm to replace the old. Then the science can return to a paradigmatic state. A classic example was when classical Newtonian mechanics was replaced by Einstein's relativity theory. Yet because the scientific revolutionaries are creating their ideas without the same paradigmatic restraints as those practicing normal science, might it not be the case that the former would be at higher risk for mental illness? After all, during the crisis phase the natural science has in a sense become more comparable to a social science.

An empirical study conducted by Young-gun Ko and Jin-young Kim provides some support for this conjecture. Starting with 76 scientific geniuses, the researchers divided them into four groups according to their degree of psychopathology: 22 with no psychopathology, 27 with personality disorders, 13 with mood disorders, and 14 with schizophrenia, which was inclusively defined to encompass other psychotic disorders. In addition, these same geniuses were assessed on achieved eminence and the contribution on which their eminence was based—whether paradigm-preserving contributions or paradigm-rejecting contributions. Paradigm-preserving contributors were like Kuhn's "normal" scientists whereas paradigm-rejecting contributors were akin to Kuhn's scientific revolutionaries. These three sets of variables—psychopathology, eminence, and contribution type—all interacted in a very fascinating manner. Those scientific geniuses who exhibited no psychopathology were more likely to become famous for paradigm-conserving contributions, whereas those geniuses who exhibited some degree of psychopathology were more likely to become famous for paradigm-rejecting contributions. Furthermore, the latter effect was most pronounced for those with schizophrenia or other psychoses.

Isaac Newton certainly illustrates the last trend. Although he was supremely eminent in his own time and remains so in our own, he is also viewed as a major scientific revolutionary, completely overthrowing the Cartesian paradigm that prevailed in his day. Even so, he suffered from a serious and multifaceted psychopathology that included symptoms of bipolar disorder, autism, and paranoid schizophrenia. These symptoms are not just conjectural, for their presence is overtly manifested in his correspondence. He once sent a rather paranoid letter to the philosopher John Locke accusing his friend of trying to ensnare him with women—this when Newton was a lifelong bachelor and misogynist who most likely died a virgin. He was not a paragon of psychological well-being.

Parallel Fractal Patterns in the Arts

Ko and Kim's investigation was inspired by Ludwig's earlier attempt to explain why the risk of mental illness varies according to domain of achievement. Not just the difference between scientific and artistic genius, but also contrasts among the sciences and the arts seem to betray a regularity. Specifically, Ludwig hypothesized that "persons in professions that require more logical, objective, and formal forms of expression tend be more emotionally stable than those in professions that require more intuitive, subjective, and emotive forms." Hence, the contrast between normal and revolutionary geniuses within the natural sciences could merely represent a special case of this tendency—much like the parallel difference between natural and social sciences. Ludwig made even finer differentiations within artistic creativity:

• Geniuses in the formal arts (such as architecture, design, and composing) exhibit less psychopathology than those in the performing arts (such as singing, dancing, acting, and directing), who in turn exhibit less psychopathology than those in the expressive arts (namely, the visual and literary arts).

• Among literary geniuses, poets are more at risk for psychopathology than are fiction writers (both novelists and story writers), who in their turn are more at risk than nonfiction writers (such as biography, literary criticism, and journalism). The contrasts among these three genres are especially conspicuous for depression—hence the prominence of suicidal poets.

• Artistic geniuses working in a highly emotive style (expressionism, art brut, and abstract expressionism, for example) should exceed the psychopathology seen in those working in a symbolic style or movement (such as realism, naturalism, impressionism, photorealism, surrealism, Dada, social realism, postmodern realism, postimpressionism, pop art, and the Vienna Secession), while the "symbolic-style" artists should exceed the psychopathology of those working in a formal style (abstract, futurism, constructivism, fauvism, cubism, op art, suprematism, and conceptual art). The ordered contrasts among these three styles are manifested in alcoholism, drug abuse, depression, mania, and anxiety, among other symptoms.

According to Ludwig, these regularities represent a fractal pattern of self-similarity in which the parts echo the whole—a metaphor well worth contemplating!

Not only was Ludwig able to show that this predicted pattern corresponds with his own data, including those geniuses seen in table 2.1, but other empirical studies have found similar patterns. Poets especially are at the highest risk. Apparently creative genius must find the right fit between any psychopathic proclivities and the niche defined by a particular creative domain. This need for fit implies that a person's specific tendencies toward mental illness might be either too little or too much with respect to a given domain—like the fable of Goldilocks and the Three Bears. Hold onto that thought: it will receive reinforcement very soon!

Superlative Genius versus Also-Ran Creators

As we begin to look at whether the risk of psychopathology changes with the degree of creative achievement, it's important to repeat that the conclusions from the first section on how creative geniuses differ from the rest of us are unrelated to the conclusions from the previous section on how mental illness varies across types of creative achievement. Yet they both represent legitimate ways of addressing the issue of whether genius is related to madness. We first found that for most creative domains, the lifetime rates for psychopathology exceeded the crude baseline set at 50%. We then saw that the specific rates varied across the domains so much that in some domains, such as the natural sciences, the percentages actually fell below baseline. The net result is that scientific geniuses as a group display higher mental health than the norm, whereas artistic geniuses as a group tend to exhibit lower mental health than the norm. "Crazy" artists are more common than "mad" scientists.

But now conduct this simple mental experiment: Suppose we reduce every one of the rates for all domains of genius to *half* their empirical value. Just multiply each percentage by 0.5. Then the second question remains true, but the first question becomes instantly and undeniably false. Creative geniuses would now exhibit *less* mental illness than everybody else despite substantial contrasts among geniuses in scientific and artistic domains. Even taking the extreme case, the 87% rate for poets would shrink to about 44%, which falls just below baseline expectation—but still *three* times higher than the new 14% rate recalculated for the natural scientists! The conclusion? Contrary to what some skeptics have argued, the mad-genius questions—note the plural—remain critical to our understanding of creative genius. Anyone who argues otherwise just hasn't done the math.

The third and final question, then—whether the greatest geniuses are even more prone to psychopathology than are far more obscure creators contributing to the same domain—is also orthogonal to the previous two. Ko and Kim already introduced achieved eminence when they compared paradigm-preserving scientists with paradigm-rejecting scientists, yet here we want to look at a broader range of creative domains. And we need to introduce two complexities, one logical and the other empirical.

Logical Complexity: Great Geniuses Go Mad while Most Creators Are Perfectly Sane

I mentioned the Positive Psychology movement earlier. Beginning at the onset of the current millennium, positive psychologists endeavored to shift the focus of scientific research from the negative to the positive—including from mental illness to mental health. Mental health was deemed associated with many other human strengths and virtues, including genius and creativity. Not surprisingly, many took issue with the mad-genius idea, believing it yet another example of negativistic thinking. In support of their Pollyannaish position was a body of research; it seemed to show that creativity was negatively correlated with psychopathology. Yet given that this research did not directly involve creative geniuses, as I observed earlier, it can't really tell us anything definite about the mental health of those who have made the biggest contributions to the arts and sciences. In fact, it is logically feasible for creativity and psychopathology to positively correlate even when creative people as a whole are far less prone to psychopathology than are noncreative people. This seeming contradiction has been styled the "mad-genius paradox."

Two Fundamental Observations on the Range and Distribution of Genius The mad-genius paradox follows naturally from the two key features of achieved eminence as our preferred indicator of creative genius.

First, the range in eminence is immense. Take poets, for example. The 1972 edition of *Oxford Book of English Verse* lists 602 poets sufficiently important for inclusion. The psychologist Colin Martindale determined the relative fame of these poets by looking up the number of books written about each and every one according to the Harvard University Union Catalogue. Can you guess which poet attracted the most books? Correct: William Shakespeare, with a total of 9,118 volumes! Next came John Milton, with 1,280, and then Geoffrey Chaucer, with 1,096. But how low does that book count go? To zero! Some 134 poets famous enough to make it into the 602, a status many English poets could only envy, still don't manage to inspire a single literary scholar to write a book about them. Apparently not even a doctoral thesis written by some graduate student in an "English Lit" program! Hence, by this criterion, poetic fame ranges from 0 to 9,118. That's really huge!

Second, the distribution is highly elitist: a few creative geniuses hog most of the fame, while the overwhelming majority of them wallow in extreme obscurity. Returning to Martindale's study, of the total 34,516 books written about the 602 poets, the triad of Shakespeare, Milton, and Chaucer together account for 11,494, or almost exactly one-third! The top dozen poets can claim 50% and the top quarter about 65%. That means that the remaining 577 poets have to compete for the remaining 12,000 or so volumes, or about the same number that the top triad have already claimed exclusively for themselves. Very unfair! And, indeed, there are not enough books to go around, particularly when the poets just beyond the top 25 continue these monopolistic practices. At the bottom of the heap are those 134 poets, or slightly more than 22%, who get totally neglected by the literary scholars. Not a single volume dedicated to honoring their names! The supremely famous creators are not only the tip of the iceberg above the ocean surface, but the topmost of the very tip—while the very bottom of the iceberg is very vast and gloomy.

Needless to say, the picture right below the iceberg becomes even more dismal if we look beyond the English poets who made it in the *Oxford* volume. Wikipedia includes entries for more than 200 who didn't make the cut and yet are still important enough to have an online article about them. And the number of poets in the outer fringe who publish in minor literary magazines is certainly far greater still. These poets are decidedly creative—a few have even sent me their attractively designed chapbooks—but none are creative geniuses. Their poems will never make it into a critically acclaimed anthology or popular audiobook, nor will they earn an entry in a future reference work or website.

Although this example narrowly focuses on English poetry, the statistics are typical for every imaginable domain of creativity. A tiny fraction of the indubitable creative geniuses gather most of the fame while a very large proportion scurry around in darkness. Worse still, the gap between elite top and bottom is so large that the creators hardly belong to the same species. Genius is not even in the same genus.

One Fundamental Derivation: The Most Eminent Creator Is the Least Representative Why are the above two observations so critical? The creative geniuses at the top can easily display higher risk for psychopathology even when creators as a whole exhibit a lower risk for psychopathology. More specifically, the magnitude of achieved eminence can correlate positively with a propensity for mental illness despite the fact that creativity is a reliable sign of mental health. That can easily occur because the creative geniuses at the tip of the iceberg are far too few to affect the calculation of the overall risk. Shakespeare, Milton, and Chaucer may have grabbed 33% of the biographies and literary criticism, but they still represent less than half of 1% of all poets (3/602 = 0.005 = 0.5%). The more eminent the creator, the less representative they are of all creators: that logical necessity just won't go away. It also tells us that all of the psychological research conducted on samples of college undergraduates and even minor creators cannot provide a conclusive estimate of the risks for the creative geniuses residing at the apex of fame. Absolutely not!

Observe carefully, as well, that any relation between achieved eminence and psychopathology implies nothing, one way or another, about the other two questions. Beyond doubt, the risk can vary across creative domains without any regard for this third question. And the same holds for the relation between the first and third question. In particular, the lifetime propensity for mental illness for the world's greatest geniuses might be lower than 50%, and yet eminence still correlate positively with the inclination. For example, the rate might ascend from 0% for the near nonentities to 40% for the most eminent, still yielding a highly positive correlation—and the average rate across all creators would approach even if not equal 0%! Or the answers to the first and last question might work the other way, too. The illness rate might be identical for both ends of the eminence distribution, but the rate lie above or below the 50% criterion. For the last time, researchers antagonistic to the mad-genius hypothesis really have to do their math! Even more importantly, if you want to address the third mad-genius question, you have to study bona fide creative genius, period. Astronomers study the stars, not terrestrial pebbles.

That is precisely what we'll do next.

Empirical Complexities: When Creative Genius Finds the Madness Sweet Spot

Amazingly, rather little empirical research has been conducted on the third question—probably because it's too often confused with the first. If somebody shows that highly eminent creators are more likely to show higher risk for mental illness than the general population, doesn't that automatically prove that eminence correlates with psychopathology? But the answer is no, as we've just seen. Admittedly, some research using personality inventories have obtained some suggestive findings. Just consider these two instances:

• The first inquiry examined 257 German painters and artists, using art experts to single out the 60 most eminent artistic creators in the sample. The latter luminaries were found to score noticeably higher on the EPQ Psychoticism scale, a measure of subclinical symptoms of psychopathology. Though not outright ill, high scorers are not particularly pleasant people either! Who of us genuinely like persons prone to be egocentric, antisocial, tough-minded, cold, unempathetic, impersonal, aggressive, and impulsive, however creative they might be?

• The second investigation scrutinized 56 successful creative writers, subjecting them to numerous tests, including the MMPI.

Again, based on expert evaluations, these authors were split into 30 who were highly eminent and the remaining 26 who were successful but not eminent (that is, their writing helped pay the bills). Although all writers scored above the norm on the clinical scales, the eminent writers scored even higher on those scales than did their less distinguished colleagues. Indeed, creative writers fell in "the upper 15 percent of the general population on *all* measures of psychopathology furnished by this test." That means even stronger leanings on the scales of psychoticism traits toward hysteria, schizophrenia, depression, psychopathic deviation, psychasthenia, hypochondriasis, paranoia, and hypomania. You don't have to know what these terms signify to guess that eminent creative writers are by no means happy campers. But there's good news as well: the elite 30 do not score as high as clinical populations who regrettably succumb to mental disorder.

I could provide other instances besides the two inquiries above, but it's best to stop here. I promised at the outset to provide as many concrete illustrations as possible to keep the text from getting too abstract. That's not possible for psychometric research that observes confidentiality agreements. The 60 eminent artists and the 30 eminent writers are as anonymous as their comparison companions. With no names, there can be no examples.

Hence, I wish to revert to two investigations with which we are already familiar, inquiries that actually published the full list of their sampled creators. Although all of the creators are eminent to some degree, they still vary substantially in magnitude—from somewhat obscure also-rans to the world famous. The two investigations, of course, are those by Ludwig and Post.

Ludwig Revisited: Psychological "Unease" and Creative Achievement Table 2.1 displays some of the most eminent creators in Ludwig's study, but what about the other end? How big is the spread from bottom to top? Well, here are some of his lesser creators: the poet Charlotte Mew, the playwright Stanisława Przybyszewska, the fiction writer Evelyn Scott, the painter Charles Schreyvogel, the scientist St. George Jackson Mivart, and the theologian Lambert Beauduin. If you know who they are, you did better than I did. I was forced to google their names.

Interestingly, instead of devising a measure of achieved eminence, Ludwig decided to assess creativity using the far more elaborate and sophisticated Creative Achievement Scale (CAS). This scale applies specifically to the works on which a person stakes his or her enduring reputation. It includes such criteria as these five:

1. "Are creations, products, performances, or works likely to be appreciated long after person's era even though the person's actual name may not be remembered ...?" For example, many of the photographs that Ansel Adams took of Yosemite Valley have become iconic representations even if a viewer might not remember who actually snapped the photo. And innumerable people throughout the world are thoroughly familiar with the sci-fi concept of the "time machine" without being aware that the term was coined and illustrated in H. G. Wells's novel by the same name.

2. "Did personal product, ideas, or work have broad human application, apply to Western civilization in general, or embody universal values or ideals?" For instance, Jean-Paul Sartre's literary and philosophical contributions not only won him a Nobel Prize for Literature—which he brazenly turned down!—but also

played a signal role in defining existentialism, the modern philosophy with perhaps the strongest implications for everyday life.

3. "Did person rise above limitations of his or her society or era by setting new directions, anticipating social needs or foreseeing future?" I'd place Alan Turing's prescient ruminations about artificial intelligence in this category, for great thinkers today still grapple with the issues he raised in the mid-twentieth century. Have you seen the 2015 film *Ex Machina*? Or heard Stephen Hawking's ominous warning about AI ending the human race?

4. "How influential was person on contemporary and subsequent professionals (protégés, disciples, adherents)?" What about Niels Bohr, a Nobel laureate who helped train later Nobel laureates, such as Harold Urey, Wolfgang Pauli, Linus Pauling, Werner Heisenberg, Felix Bloch, Max Delbrück, and Lev Landau (not even counting his immortal debates with Albert Einstein over quantum theory).

5. "How original was the person's main work, product, or accomplishment?" This is easy. Just think of a big name, and at least one highly original achievement should spring immediately to mind: Marie Curie, radium and polonium; Bertrand Russell and Alfred North Whitehead, *Principia Mathematica*; James Joyce, *Finnegan's Wake*; T. S. Eliot, *The Waste Land*; Pablo Picasso, *Guernica*; Auguste Rodin, *The Thinker*; Igor Stravinsky, *The Rite of Spring*; and so on.

All told there were 11 items of variable importance used to calculate a summary CAS score. All of those creative geniuses shown in table 2.1 scored in the top quartile on this measure, whereas Mew, Przybyszewska, Scott, Schreyvogel, Mivart, and Beauduin all scored in the bottom quartile. Seem reasonable?

Ludwig then showed that these CAS scores correlated positively with his indicators of psychopathology. More creativity, more risk of psychopathology—yet with an upper limit. Clearly, lifelong debilitating mental illness would most likely yield CAS scores of zero. He then averred that "the presence of psychological 'unease,' potentially but not necessarily produced by any mental illness that is not too incapacitating, contributes to the realization of true greatness"—or what might also be called genius. At the same time, Ludwig added that other crucial traits should accompany this "unease" to moderate excessively adverse effects. This qualification will receive attention later, but to provide a teaser to keep you reading, part of Tip 1 has an intimate connection with Tip 2.

At this point, a reader might object: Ludwig used scores on *creative achievement* rather than *achieved eminence*, the strict object of the third question. True, but the two correlate highly with each other, even if they can't be considered equivalent. Besides, isn't a positive correlation with creative achievement just as important as one with achieved eminence? Either correlation supports conjectures about the mad genius. Hence, the substitution inspires further confidence in the inference, not less.

Post Post-Analyzed: Enjoying Just the Right Amount of Psychopathology Although Post was quite meticulous about assessing the magnitude of subclinical psychopathology in his creative geniuses, he made no effort to gauge their degree of genius, whether by achieved eminence or creative achievement. Fortunately, a recent follow-up remedied this neglect, taking advantage of independent and highly reliable measures of achieved eminence independently compiled by another researcher. After converting Post's mental illness scale into numerical form, it then became possible to scrutinize the relation between psychopathology and eminence. Furthermore, because we already know that the relationship may depend on the domain of creativity, and that the relationship may not be linear, allowance was made for curvilinear, single-peaked functions. Figure 2.1 shows the outcome. The results are thought provoking.

To start, of the five domains represented, only the writers and artists display consistently positive functions. Going from none (0) to severe (3) corresponds to a steady increase in expected achieved eminence. The result certainly provides a complement to the higher risk rates for these two groups noted earlier.

In contrast, the other three domains reveal the presence of a sweet spot, albeit its exact location depends on the domain. For the thinkers, the optimum falls somewhere between marked (2) and severe (3), whereas for the composers the peak lands

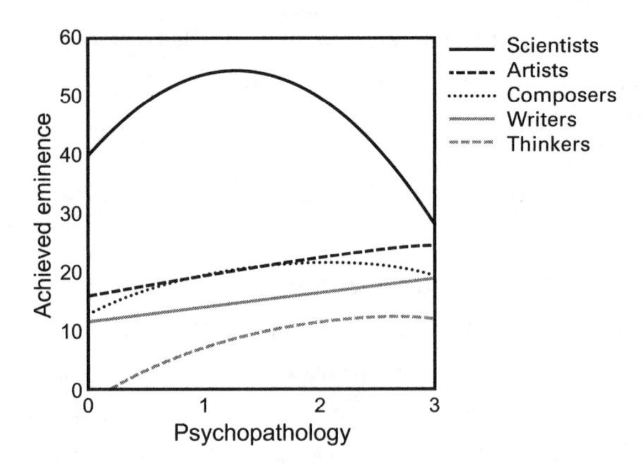

Figure 2.1
Curves describing the relation between achieved eminence and degree of psychopathology (0 = none, 1 = mild, 2 = marked, and 3 = severe) for five separate groups of creative genius.
Source: Adapted from Simonton 2014c, p. 58.

roughly between mild (1) and marked (2). That said, the overall tendency of these two curves remains upward, so those with severe psychopathology are still higher in acclaim than those with no psychopathology whatsoever. That tendency is not seen for the scientists, however. Quite the contrary! Those with severe psychopathology are less eminent than those with none. In addition, although a sweet spot occurs between mild and marked, it's somewhat closer to the mild end of the spectrum in comparison to the composers, thus minimizing all the more the place of mental illness in scientific genius.

These three roughly "inverted-U" curves help settle an issue sometimes raised against this research. Perhaps the more eminent creators are more likely to show psychopathological symptoms because they are more likely the subjects of longer biographies that can provide more room for episodes recounting such events. Yet that's inconsistent with the finding that the most eminent scientists, thinkers, and composers actually exhibit less mental illness than those somewhat less eminent. Indeed, the scientists with no psychopathology are more eminent than those with severe psychopathology. How can that be if this curve represented a bias of biographers?

Combining these results with the earlier findings of Ludwig as well as Ko and Kim, it becomes clear that achieved eminence *is* associated with mental illness. So the broad-stroke answer to the third question driving this chapter is a resounding "yes!" Even so, the precise nature of that response is far more intricate than anybody ever imagined. Sometimes the relation can be positive, other times negative, and still other times curvilinear, with the specific location of the optima varying according to creative domain. No one-size-fits-all response is even possible.

Just to prove conclusively how divergent the relations can be, do you know what happens when you try to fit a single curve to merge the five separate curves in figure 2.1 into a single summary curve? You get absolutely nothing! The diverse relationships cancel out, yielding zip. No positive or negative relations, no linear or curvilinear functions—just a pristine null effect. Therefore, answering the third and last question is not as easy as first meets the eye. Yet who said that the science of genius had to come up with simple answers? After all, the phenomenon just may be really, really complicated!

The Paradoxical Upshot

Back in 1681, the English poet John Dryden penned "Great Wits are sure to Madness near ally'd, / And thin Partitions do their Bounds divide." Hence, genius-level creativity is not equivalent to psychopathology. Not only are there numerous creative geniuses who show no signs whatsoever of mental illness, but there are far, far more mentally ill persons who show no signs whatsoever of creativity. So where's the partition?

As I've already cautioned, mental illness is not a discrete all-or-none state. Rather, it consists of multiple symptoms that can vary in frequency and intensity. With respect to frequency, all it takes is one major episode in an entire lifetime to enter the reported statistics. We're not talking about a disorder so frequent that it effectively lasts a whole lifetime. A serious emotional breakdown or suicide attempt should do. That means that every creative genius will have more than a sufficient number of "lucid moments" for productive work at the bench, desk, easel, or keyboard. And regarding intensity, we have already seen how often psychopathological symptoms operate at subclinical levels. This was true even for many if not most of those who Post placed in the severe category. For example, the Austrian composer Anton Bruckner may have endured a severe obsessive-compulsive disorder, but at least he often put it to good use. Besides compulsively counting things, such as the measures and rhythms in his symphonic compositions, he would also revise his works so obsessively as to introduce the "Bruckner Problem"—how to identify the best versions to perform and record today! Yet he kept on composing until his death. Excessive perfectionism did not paralyze his creativity—nor, for that matter, did his rather morbid obsession with dead people.

But another aspect of the thin partition remains to be discussed: the impact of general intelligence. Although creative geniuses share many traits with those who are mentally ill, they are also bright enough to ameliorate the negative effects, and even exploit the positive. This possibility is well documented in research using the Creative Achievement Questionnaire (CAQ; not to be confused with the CAS discussed earlier). The CAQ asks respondents to indicate their creative achievements in several different domains, such as science, invention, creative writing, theater and film, music, dance, architecture, and the visual arts (painting and sculpture). Achievements in each domain are scaled from absent to very high levels. For instance, the scale for the visual arts runs as follows:

0. I have no training or recognized talent in this area.
1. I have taken lessons in this area.
2. People have commented on my talent in this area.
3. I have won a prize or prizes at a juried art show.

4. I have had a showing of my work in a gallery.
5. I have sold a piece of my work.
6. My work has been critiqued in local publications.
7. My work has been critiqued in national publications.

Although the scales may stop just short of true creative genius, they at least trace the pathway that's often taken to the door's threshold.

CAQ scores also correlate positively with performance on several tests designed to measure either creative thought or the creative personality, such as divergent thinking and openness to experience. But another correlation is even more significant: a positive link with *cognitive disinhibition*—the failure to filter out extraneous stimulation. Lacking a highly disciplined filter, the person's consciousness is often bombarded with what should be ignored. From the standpoint of creativity, cognitive disinhibition would seem a tremendous asset. It enables someone to notice novelties that others might miss, to make connections between ideas that others might overlook. A celebrated example occurred when Alexander Fleming noticed that a blue mold was ruining his bacteria culture. Most scientists would just have tossed the petri dish into the autoclave for later use, totally ignoring the potential implications. But instead, Fleming investigated the antibiotic agent the mold was emitting to wipe out the staphylococci. The end result was the development of penicillin and Fleming's eventual receipt of the Nobel Prize for Physiology or Medicine.

Unfortunately, cognitive disinhibition has a downside: it's a common symptom of mental illness. Many psychoses, such as schizophrenia, involve a person becoming simply overwhelmed by external events and internal thoughts. The net result are visual and auditory hallucinations as well as delusions, including the delusions of grandeur and persecution underlying paranoia. These are what John Nash suffered when he lapsed into in a psychotic episode. Thus, the same cognitive disinhibition that nurtures creativity can also undermine a person's sanity. So how does someone take advantage without the devastating disadvantages?

It turns out that there are several abilities and personality traits that help convert a vulnerability to mental illness into a potential for creative genius. Yet perhaps the best buffer is high general intelligence—even that measured by an IQ test! Someone with a really high IQ would be able to handle the increased influx of extraneous information, and convert the added input into original ideas. The potent intellect would then transform a clinical deficit to a subclinical asset. As Dryden might say, a high IQ inserts the thin partition between the crazy and the creative. Naturally, you're still not required to take an IQ test to make this happen, but just enjoy the intellectual ability to do well on the test if you did take it. And the bigger the onslaught produced by cognitive disinhibition, the better your general intelligence probably ought to be.

I discovered online that Salvador Dalí, the famed Spanish painter, could boast an IQ of 180. I have no idea where this superlative score came from. But given the bizarre images that float in his wildest surrealistic paintings, it seems like IQ 180 was just barely high enough to keep him sane. That IQ partition would then explain Dalí's own perplexing paradox: "The only difference between me and a madman is that I'm not mad."

Tip 3
Start Out as a Zygote with Super Genes / Carefully Pick Your Home and School!

The science of genius began with Francis Galton, who was *both* a genius scientist and a scientific genius. We already know one reason why he might be called a genius, for I mentioned when discussing Tip 1 that his IQ was estimated at close to 200. He was certainly a scientific genius, too, because he made highly creative contributions to psychology, anthropology, geography, meteorology, criminology, statistics, psychometrics, and genetics—a bona fide polymath in science. To provide a mere glimpse of his scientific achievements, Galton made a name for himself as an explorer in uncharted southwest Africa, introduced weather maps and scientific forecasting, devised methods for identifying criminals using finger prints, ventured the first attempt to measure human intelligence, and pioneered statistical techniques that became the mainstay of the biological and behavioral sciences. He even invented the dog whistle to study auditory acuity in the high frequencies! Unfortunately, his creative genius also had a dark side, for he coined the term *eugenics* and thereby promoted the selective reproduction of the brightest and the best among *Homo sapiens*—an idea that became distorted to horrific purposes by the Nazis during World War II. Ironically, Galton didn't practice what he preached: Despite his own "good genes," his 43-year marriage produced no children!

In 1869 Galton published his first book entirely devoted to the scientific study of genius. Because it was also the very first book exclusively dedicated to the science of genius, it deserves its own section for full discussion.

Hereditary Genius

In *Hereditary Genius: An Inquiry into Its Laws and Consequences,* Galton attempted to prove beyond any shadow of doubt that general intelligence was strongly inherited from parent to child. Beginning with the assumption that exceptional intelligence was virtually equivalent to achieved eminence—and thus implicitly obliterating the paradox suggested in Tip 1—he then predicted that genius should run in families.

To test this prediction, he collected biographical data on geniuses in a diversity of achievement domains, such as scientists, poets, composers, painters, commanders, and politicians. He then looked for distinguished family pedigrees—clusters of geniuses extending two or more generations with some genetic relationship. A particularly remarkable example appears in classical music: the Bach family. Although the German composer Johann Sebastian Bach is by far the most famous member, his lineage dates back to a Hans Bach who died in 1626, more than a century before J. S. Bach's death in 1750. All told, more than 20 Bachs could be considered eminent musicians to some degree. Without question, the most famous composers in the family, besides J. S. himself, are his sons Carl Philipp Emanuel (C. P. E. or the "Berlin Bach"), Johann Christian ("London Bach"), Wilhelm Friedemann ("Halle Bach"), and Johann Christoph Friedrich

("Bückeburg Bach"). Of course, not all of these Bachs were equal in musical genius. C. P. E. Bach's fame as a composer for a time eclipsed that of his father. This was the Bach who would first come to mind when Franz Joseph Haydn or Mozart used that surname. In contrast, Wilhelm Friedemann, though clearly talented, claimed a much less successful career, and died in poverty. So remarkable is this sibling pedigree that the musical satirist and self-proclaimed "Professor" Peter Schickele couldn't resist creating the fictitious P. D. Q. Bach, the "21st child" of J. S. Bach (who only had 20 children) and the latter's "only forgotten son." Yet despite his nonexistence, P. D. Q. somehow managed to compose dozens of works, including the infamous *1712 Overture* and *The Abduction of Figaro*!

Because Galton was strongly orientated toward quantification—he had been a Cambridge University math major—he subjected these data to statistical analyses. He concluded that the familial clustering of genius far exceeds what would be expected by chance alone (especially given the extremely low base rate of genius in the general population). Furthermore, for the most part, the closer the family relationship the higher the odds that genius might be shared. To be sure, he noted some quirks in his analysis. For instance, he found that the inheritance of musical genius only took place through the male lines. That would seem to defy a genetic explanation—unless hereditary genius can operate like sex-linked traits. A "genius as color blindness" model? Even so, Galton started his research at about the same time that Gregor Mendel began studying the inheritance of traits in peas, and Mendel's momentous discoveries were largely unknown until 1900. Nobody knew back then how traits actually transferred from one generation to the next. Creative genius is certainly not a gene on the Y chromosome!

There's something obliquely self-serving about Galton's *Hereditary Genius*. In his chapter on scientific genius, he traces the distinguished pedigree of the Darwin family. For example, the grandfather of Charles Darwin, the eminent English naturalist, was Erasmus Darwin, who authored an early version of evolutionary theory long before his grandson's landmark *The Origin of Species*. Galton closes discussion of this lineage with the following cryptic assertion: "I could add the names of others of the family who, in a lesser but yet decided degree, have shown a taste for subjects of natural history." Who might he be referring to? Why, to some of Galton's close relatives and maybe even to Galton himself (given that his approach to studying *Homo sapiens* was often not unlike what's done in natural history). To cut to the chase, Erasmus Darwin was also his grandfather, though via a different grandmother—but what did that matter given what Galton said about female lines? That makes Galton and Darwin half first cousins, albeit Darwin was the elder by more than a dozen years. Hence, Galton's book indirectly proves that he himself might have hereditary genius!

Indeed, that very proof became even stronger before Galton passed away in 1911. Four of Charles Darwin's sons would add to the distinguished family name: Sir George Howard Darwin, a Cambridge astronomer who formulated the fission theory of our moon's formation from the primeval earth; Sir Francis Darwin, a botanist who conducted experiments on phototropism with his father; Sir Horace Darwin, a distinguished civil engineer; and Major Leonard Darwin, the only one of the four who was neither knighted nor elected Fellow of the Royal Society (or FRS; roughly comparable at the time to winning a Nobel Prize). Although Leonard considered himself less brilliant than George, Francis, and Horace, he still had a highly accomplished career, succeeding

Galton as head of the British Eugenics Society and serving as the mentor of Ronald Fisher, the noteworthy evolutionary biologist and statistician (after whom the ubiquitous "F Test" in data analysis is named). Charles Darwin's four sons were chips off the old block just like J. S. Bach's were. Plus, all four were Galton's half second cousins, thus strengthening his claim to genius genes!

Galton thought that his 1869 book established once and for all the case that genius was born, not made. Yet that's not the end of the story.

Nature ... or Nurture?

You'd think that anyone who found himself explicitly included in one of Galton's illustrious pedigrees would feel so flattered that they'd automatically endorse his genetic theory of genius. But that's not what inevitably happened. Galton had noted that Augustin Pyramus de Candolle, an eminent Swiss botanist, had fathered the distinguished botanist Alphonse Pyramus de Candolle. Indeed, although nobody could know it at the time, Augustin's son, grandson (Anne Casimir Pyrame de Candolle), and great grandson (Richard Émile Augustin de Candolle) would all become notable Swiss botanists, forming a four-generation dynasty! Yet Alphonse wasn't buying into Galton's biological explanation. Instead, certain environmental factors must stimulate the emergence of genius, which was thus made, not born. To support his counterargument, Alphonse collected data on the circumstances that support the emergence of scientific genius. His special focus was on why certain nations at specific times seem most productive of great scientists. He published his findings at length in 1873.

Galton was not to be outdone, so he immediately put together an empirical study of his own in response. Because he had been elected FRS, he chose to survey his fellow Fellows with a questionnaire asking them about their personal backgrounds. That means that his respondents included some of the top scientific geniuses residing in Great Britain at that time, among them the mathematician Arthur Cayley, the physicist James Clerk Maxwell, the astronomer William Lassell, the mineralogist Nevil Story Maskelyne, and the biologists Thomas H. Huxley, Richard Owen, and Charles Darwin. This survey was also the first self-report measure of its kind. Better yet, more than 100 scientists responded. Darwin's survey responses actually survive to this very day; his son Francis reproduced them in his father's *Life and Letters*.

Even more crucial was his innovation regarding how to frame the questions addressed by the survey. In particular, he explicitly defined the nature-nurture issue: "The phrase 'nature and nurture' is a convenient jingle of words, for it separates under two distinct heads the innumerable elements of which personality is composed. Nature is all that a man brings with himself into the world; nurture is every influence from without that affects him after his birth." Although the alliterative use of nature and nurture can be found in Shakespeare's *The Tempest*—when Prospero complains about his vain efforts to civilize Caliban—Galton was the first to introduce the nature-nurture issue as a fundamental scientific question. And to emphasize its importance, the resulting 1874 book was given the title *English Men of Science: Their Nature and Nurture*. Yes, it was published only one year after Candolle's book. Fast work!

Although Galton hadn't given up his commitment to hereditary genius, he became willing to allow some environmental influences. For the latter, his questionnaire concentrated on two that have received a great deal of subsequent research: family background and educational experiences. After looking at the most telling findings of that research, I will return to the nature side of the question before combining the two.

Genius Made? Parents, Teachers, and Mentors

Let us examine more closely the illustrious Darwin (and collateral Galton) family "pedigree" in *Hereditary Genius* in the context of the nurture option presented in *English Men of Science*. Whatever genes may have been passed down the line, the Darwins enjoyed familial and educational opportunities that the average British citizen could only envy. Charles's father was Robert Darwin, a popular physician with the wealthy and a shrewd investor who accumulated a substantial fortune. Robert had begun his medical studies at the prestigious University of Edinburgh before his father, the famed Erasmus, sent him to earn his MD at Leiden University, the oldest university in the Netherlands. At only age 20, Robert made an important scientific discovery, providing the first empirical evidence for the eyes' microsaccades (the small, involuntary, jerk-like movements made while trying to fix the vision on a particular object). He published his findings in the *Philosophical Transactions of the Royal Society*, and was elected FRS two years later. After acquiring his substantial reputation and income, Robert married the favorite daughter of Josiah Wedgwood, his father's close friend as well as the most renowned pottery manufacturer of his day—the creator of classic vases, teapots, cups, and plates still proudly exhibited in museums all over the world. Robert and Susannah had six children, among them Charles.

Needless to say, Robert was able to send his (two male) children to top-tier universities where they could take classes with eminent teachers and work under prominent mentors. Most particularly, Charles learned his geology and botany under the illustrious Cambridge professors Adam Sedgwick and John Stevens Henslow. It was the latter botanist who recommended his protégé for the life-changing position as unpaid naturalist aboard the *Beagle*, which was about to circumnavigate the world, including exploration of the Galápagos Islands. The recent college graduate was thereby provided with a wealth of unique geological and biological knowledge that helped prepare him to author *The Origin of Species*. Doesn't it seem that Charles was more made than born? What would have happened to his "innate" creative genius if Robert had refused to let Charles go, which was his father's first reaction before being persuaded otherwise by his brother-in-law Josiah Wedgwood II?

Empirical research shows that Charles's upbringing, education, and training are not atypical, at least not for scientific geniuses.

Socioeconomic Background In Galton's 1874 survey, almost a third of those elected FRS came from professional families where the breadwinner was a lawyer, physician, minister, or teacher. Another two-fifths grew up in the homes of bankers, merchants, and manufacturers. The remaining quarter or so emerged largely from families headed by fathers in the military or civil services. There was only one FRS who was the son of a farmer, and only nine who had fathers who were aristocrats or gentlemen (in the old sense of "gentry"). Galton himself had a father who was a highly successful banker. Like his half cousin Darwin, Galton was left with an inheritance sufficiently generous that he was never obliged to get a regular job.

About eight decades later Anne Roe published her important *The Making of a Scientist* in which she interviewed 64 eminent scientists. Her sample contained such Nobel laureates as George Beadle, Edwin McMillan, Hermann Joseph Muller, Robert S. Mulliken, John Howard Northrop, Linus Pauling, and Julian Schwinger. Included as well were illustrious psychologists like J. P. Guilford, Harry Harlow, Carl Rogers, and B. F. Skinner. Curiously, also in the sample was one Termite (IQ>140), Robert R. Sears, and one Termite reject, Luis Walter Alvarez (IQ<140), both discussed under Tip 1. In any event, Roe found that since Galton's day a shift had taken place: eminent scientists are more likely to come from the families of professionals—physicians, lawyers, ministers, engineers, and academics—and less likely from the families of parents involved in business, and certainly not in farming or skilled labor. Many subsequent studies have come up with the same results: most often more than half of eminent scientists originate in homes where one or both parents are professionals—a proportion far in excess of what would be expected in the general population. Even among Terman's Termites, only about a third came from such families. In this respect, Darwin's family background is more typical of modern scientists than is Galton's.

Educational Background Galton's treatment of education and training is rather deficient in comparison to what he has to say about the home environment—and more about the latter later—but he does note that "one-third of those who sent replies have been educated at Oxford or Cambridge, one-third at Scotch, Irish, or London universities, and the remaining third at no university at all." Of course, the portmanteau "Oxbridge" represents the two most distinguished universities in England, and the Edinburgh (Scotland), Trinity College Dublin (Ireland), and perhaps University College (London) were likely not too far behind. Beyond that, we must look at more recent research to show how Darwin's education must have assuredly nurtured his genius. And, in point of fact, scientific geniuses are most likely to graduate from the most elite universities.

This greater likelihood is amply illustrated in Harriet Zuckerman's study of 92 male scientists active in the United States who conducted research that earned the Nobel Prize. At the undergraduate level, "fifteen elite schools accounted for 59 percent of the American laureates but, at a roughly comparable time, produced only 12 percent of all male graduates." The elitist education continued in graduate school, for unlike in Darwin's and Galton's day, advanced training has now become mandatory for developing scientific creativity. Just "thirteen elite universities granted degrees to… 85 percent of the laureates." In contrast, the same universities produced only 55 of the degrees received by undistinguished scientists who went through graduate training at roughly the same time. Can you guess what universities we're talking about here? Yes, for undergraduate education, the Ivy League schools like Harvard, Yale, Dartmouth, Cornell, and Columbia, joined by other elite colleges like Cal Tech, MIT, Berkeley, and Chicago; and for graduate instruction, such top-notch research universities as Harvard, Columbia, Berkeley, Johns Hopkins, Princeton, and Chicago.

Naturally, because elite universities feature elite faculty, admission into such schools should increase the odds that a student will acquire a distinguished mentor, just as happened to Charles Darwin when he went to Cambridge. Acquisition is not guaranteed, of course, given that faculty even in such highly meritorious institutions can still be quite heterogeneous regarding merit. Not all are stars. Yet it remains true that creative genius tends to receive guidance under exceptional mentors. Returning to Zuckerman's Nobel laureates, for instance, "it is striking that more than half (forty-eight) of the ninety-two laureates ... had worked either as students, postdoctorates, or junior collaborators under older Nobel laureates." These master-apprentice relationships can then ramify over generations. To illustrate, figure 3.1 shows the repercussions of Lord Rayleigh mentoring J. J. Thompson, and from the latter ... well, the single picture is worth at least a few dozen words.

Take special note that each and every one of those shown in the graph received a Nobel Prize in Physics. Notice, too, that this laureate lineage explicitly involves nurture, not nature. No biological offspring whatsoever.

Need I add that comparable mentoring relations are evident in other guises of creative genius? In combination with what was said with respect to parents and teachers, hasn't a strong case been made for nurture over nature? Alas, not quite!

Genius Born? Spermatozoa, Ova, and Zygotes

The family pedigrees that Galton introduced in 1869 have been replicated in later samples more than once, so there's no doubt that they exist. Indeed, such genetic lineages can even be found among Nobel laureates. Besides the two brothers Jan Tinbergen and Nikolaas Tinbergen (separately in 1969 and 1973), there have been six father-son pairs, namely: William Bragg and Lawrence Bragg (together in 1915), Niels Bohr and Aage N. Bohr (separately in 1922 and 1975), Hans von Euler-Chelpin and Ulf von Euler (separately in 1929 and 1970), Arthur Kornberg and Roger D. Kornberg (separately in 1959 and 2006), Manne Siegbahn and Kai M. Siegbahn (separately in 1924 and 1981), and J. J. Thomson and George Paget Thomson (separately in 1906 and 1937). In addition, Irène Joliot-Curie, the daughter of Pierre

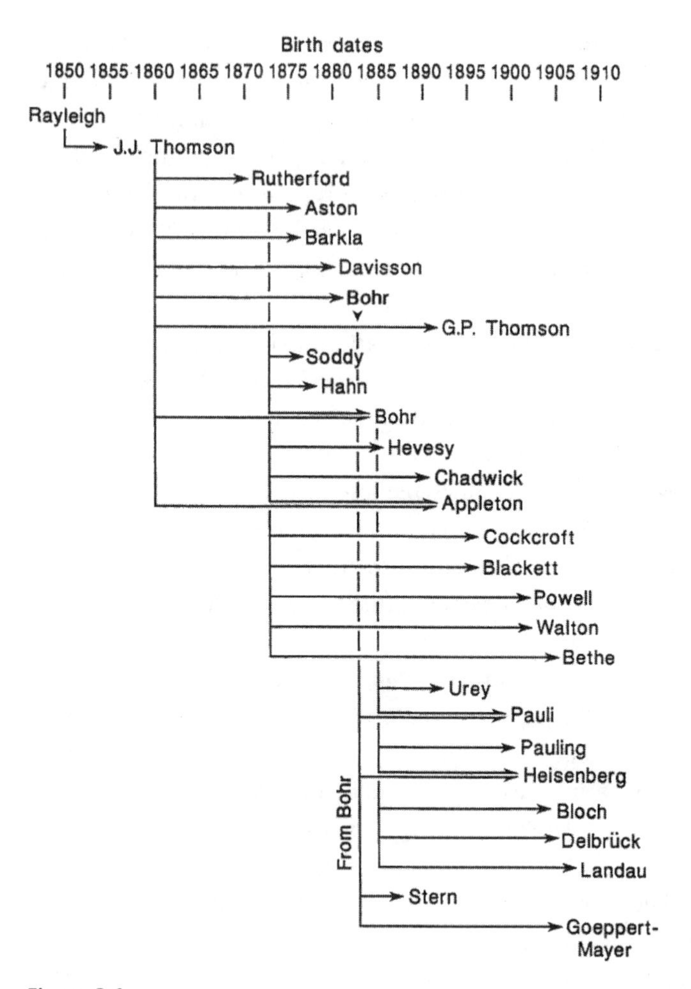

Figure 3.1
Master-to-apprentice influences among some Nobel laureates in Physics. *Source*: Simonton 1994, p. 383 (using data from Zuckerman 1977, p. 103).

Curie and Marie Curie (who shared a Nobel in 1903 before the widowed Marie received a solo Nobel in 1911), was awarded the 1935 Nobel with her own husband Frédéric Joliot-Curie.

But even if these kinships are quite astonishing, the rarity of such honors makes their implications for resolving the nature-nurture debate patently ambiguous. For example, Irène didn't just inherit "genius genes" from her mom and dad; she grew up deeply exposed to science, continuing the work on radioactivity that was initiated by her parents. Irène even met her husband when she trained him in the radiochemical lab techniques that her parents had developed. Talk about a nurturing environment for becoming an illustrious experimental scientist! And is it any wonder that Irène and Frédéric parented two highly distinguished scientists themselves, albeit not Nobel laureates? The Curie home circumstances echoed that of the Candolle's.

Given this inherent weakness in the family pedigree method, Galton worked on other techniques that would become the mainstay of behavior genetics—a sophisticated method for separating out the relative effects of nature and nurture. One of Galton's innovations was especially critical: the study of twins both identical (monozygotic) and fraternal (dizygotic). Both types of twin usually develop in identical intrauterine and home environments. Only identical twins share the same genes, however; fraternal twins are no more similar genetically than two siblings born at different times. Later behavior geneticists added a further refinement to Galton's approach: take advantage of identical twins reared apart in separate family environments—then nature will be pitted directly against nurture. Another variation on this tactic is to look at non-twin adopted children to see whether they end up more similar to their biological or foster parents. After applying rigorous statistical analyses to the resulting data, some rather provocative results emerge. The following findings are especially pertinent to the paradox of genius.

Almost without exception, any characteristic on which human beings can vary is subject to genetic inheritance to a very substantial degree. This heritability certainly applies to those abilities and traits associated with creative genius. For example, general intelligence is strongly influenced by genetic endowment. If Robert and Susannah Darwin were, like the Curies, very smart parents—and they most likely were, given their own parentage—then their son Charles should have a high likelihood of being smart too. So would Charles's own kids, given that he, as well as his father, married into a distinguished family—and hence arises a potential genetic benefit for his sons George, Francis, Horace, and Leonard. Genes even influence personality traits, such as openness to experience, as I briefly mentioned under Tip 2. Although the degree of heritability varies according to the specific characteristic—generally higher for cognitive abilities and lower for personality traits—it is not too crude of an approximation to say that nature and nurture very often deserve roughly equal credit.

Even so, because creative genius requires so many different traits and abilities, it makes no sense to speak of a "gene for genius." Indeed, each trait or ability consists of multiple genes (in other words, all are "polygenic"), so that many thousands of genes are required to produce a genius. Moreover, given that so many specific genes must converge to yield a genius, and that the inheritance of any given set of genes is largely governed by the luck of the draw—which one of innumerable swimming sperm just so happens to fuse with a randomly available ovum to form an idiosyncratic zygote at the precise moment of conception—even siblings from the same parents are not promised the same genetic gifts. From the caprice of inheritance arises two oddities: genius pedigrees without genius and genius without genius pedigrees.

Genius Pedigrees without Genius Robert and Susannah had two sons, the elder Erasmus and the younger Charles, born five years apart. Have you ever heard of the first one? Probably not (discounting the grandfather after whom he was named). Erasmus and Charles, however, grew up in the same family and even attended the same schools and colleges. Erasmus also worked under a distinguished mentor at Cambridge (in his case the chemist James Cumming). Moreover, the two brothers would frequently engage in intellectual and cultural activities together, and remained close throughout their lives, the two dying only a year apart. Yet Erasmus was clearly headed nowhere, and at the young age of 26 was pensioned off by his wealthy father. Thus prematurely retired, Erasmus was never to accomplish anything particularly notable thereafter—except carrying on a scandalous affair with the married sister of his younger brother's future wife! While this infamous event was taking place, Charles was traveling around the world aboard the *Beagle*, on his way to future fame. Given that the sibs started out with almost exactly the same environmental advantages, Charles seems to have won the genetic lottery, whereas Erasmus was dealt a really bad hand.

Genius without Genius Pedigrees Even more surprisingly, the luck can run in an entirely different direction, 180 degrees counter to Galton's family pedigrees. Creative genius can emerge completely out of familial obscurity. This inconvenient truth is exemplified by a genius no less than the English mathematician and physicist Isaac Newton. Galton struggled mightily in

Hereditary Genius with the fact that "Newton's ancestry appear to have been in no way remarkable for intellectual ability, and there is nothing of note that I can find out among his descendants, except what may be inferred from the fact that the two Huttons were connected with him in some unknown way, through the maternal line." Those Huttons were Charles, a mathematician, and James, the much more famous geologist. Yet the lineage is very tenuous, supposedly operating via a sister of Newton's mother, and with no creators of note in an intervening period that spanned nearly a century. It is telling that the Wikipedia biographies for both Huttons do not even mention that they might have been the indirect descendants of the childless Newton. If it were well documented, then you'd think it would be mentioned as a tribute to either. Hence, if Newton was born a genius, it was not by the same genetic means that Darwin might have been born a genius. Not a lot of great genes were provided in the parental hopper, but Newton somehow managed to get all the right ones. His genetic endowment then allowed him to overcome various environmental obstacles, including his mother's failed attempt to raise her firstborn son as a *farmer*. We must thank fate that nature overruled nurture. Imagine Newton in overalls growing apples instead of watching one fall from an apple tree and thereby discovering gravity!

Nor is Newton the sole instance of a creative genius sans a stellar family pedigree. Another amazing example is the great Italian sculptor, painter, and architect Michelangelo, an artistic genius quite comparable to Newton's scientific genius. Perhaps Galton was so embarrassed by this disconfirming case that he completely failed to mention the master—despite providing lineages for many other outstanding Renaissance artists, such as the Bellini family. You know who else is conspicuously absent?

An artist and scientist named Leonardo da Vinci, the natural son of a notary and a peasant—no artistic pedigree there either!

In any case, genetic effects are not always what they seem. I'll provide more niceties later. But first I must turn from nature to nurture—to discuss when nurture succeeds by failing to nurture.

Scientific Revolutionaries, Creative Artists, and Diversifying Experiences

The German composer Ludwig van Beethoven's father was a physically abusive alcoholic; although Ludwig was much closer to his mother, she unfortunately died while he was still in his teens. The French philosopher René Descartes lost his own mother when he was only 1 year old, and he ended up being raised by relatives; a very sickly child, he entered school late, and was even allowed to sleep in by the Jesuit fathers who ran the school. When only 9, the Russian novelist Fyodor Dostoyevsky suffered his first epileptic seizure, episodes that would recur for the rest of his life and eventually appear in some of his fictional characters as well. When the Austrian neurologist Sigmund Freud was born, his parents were so poor that they lived in a single rented room over a locksmith's shop, and by the time he was 4 his father's wool business went belly up, obliging the family to move from a small village (Freiberg) to the big city (Vienna). Lord Byron, the English poet, was born with a deformed right foot that caused him considerable physical and psychological anguish; yet that did not prevent him from traveling throughout southern Europe, tragically dying while fighting in the Greek War for Independence. At age 7, Thomas Edison, the American inventor, encountered so many difficulties in elementary school that his mother decided to teach him herself at home; his own curiosity led him to become a voracious reader, enthusiastically consuming such diverse fare as Thomas Gray's poem "Elegy Written in a Country Churchyard," Edward Gibbon's *The History of the Decline and Fall of the Roman Empire*, and Newton's *Principia*—albeit the last with less comprehension because his parents were incapable of explaining the math!

What do these factoids have in common—besides all describing events that happened early in the lives of creative geniuses? The answer: all offer specific examples of *diversifying experiences*. That is, all are events or circumstances occurring in childhood, adolescence, or early adulthood that "help weaken the constraints imposed by conventional socialization," ultimately generating a person who can conceive ideas less restricted by societal conventions. These constraint-relaxing experiences can include various kinds of early adversity, such as parental loss, economic instability, physical or cognitive disabilities, geographical dislocation, and problems in school or relatively little formal education. But the capacity to think outside the box imposed by norms and customs can also be acquired via less averse experiences. Omnivorous reading provides a clear example, as illustrated by Edison. Such reading can expose the person to different times and places or facts and ideas that would be otherwise unknown. Also included is extensive traveling, living abroad, and outright emigration. Indeed, multicultural encounters are highly conducive to creative development. For instance, first- and second-generation immigrants tend to be overrepresented among the eminent. The common denominator remains the same: one way or another, the future creative genius is set along a divergent path that's unconfined by what is considered "normal" according to a given culture. The upshot is often an autonomous nonconformist with an intense desire to explore and discover.

To be sure, various diversifying experiences are not all equal in developmental impact, and to some degree various experiences are interchangeable, so that two or three small challenges might equal one big one in developmental effect. Even within a single experience, we should always weigh both intensity and frequency. A person may live abroad for a semester or a decade, for example, and in a country that does or does not require learning a new language. The impact of orphanhood, to offer another case, depends on many factors, such as the youth's age, the closeness of the parent-child relationship, the number of parents lost, and their manner of death.

Artistic versus Scientific Geniuses

Importantly, diversifying experiences can operate in a manner that closely parallels what I noted in the discussion of psychopathology under Tip 2. Here also we have to recognize the possibility of too much as well as too little. Eccentricity is one thing, but out-and-out instability is another. Furthermore, the specific placement of the optimum level is likewise contingent on the domain of creative achievement. In fact, Arnold Ludwig's earlier assertion about mental illness can be paraphrased to account for how diversifying experiences vary across domains: creators active in domains that favor more "logical, objective, and formal forms of expression" should exhibit less frequent and less intense diversifying experiences than those in domains that require more "intuitive, subjective, and emotive forms."

We definitely find this predicted pattern if you look at recipients of the Nobel Prize in the domains of physics, chemistry, and literature—ignoring "physiology or medicine" as too much of a mixed bag for precise comparisons. The physics laureates come from the most stable and conventional homes, whereas the literature laureates come from the least stable and conventional, the chemistry laureates falling roughly in the middle. For example, 28% of the physicists had fathers who were academic professionals, a figure that falls to 11% for the chemists and a measly 6% for the writers. Even more strikingly, although merely 2% of the physicists lost their fathers when they were young, such paternal orphanhood went up to 11% of the chemists and fully 17% of the literature laureates. If the diversifying experiences are expanded to include parental desertion as well as death, and financial hardship or bankruptcy, then 30% of the creative writers so qualified. In contrast, the researcher concluded that the physics laureates "seem to have remarkably uneventful lives."

Another inquiry scrutinized more than a thousand eminent 20th-century personalities, finding that the creative writers grew up in unhappy home environments, whereas the creative scientists came from much more positive home conditions. The creative scientists also enjoyed superior formal education, whereas the artistic creators had very little, with the poets being least likely to enjoy any especially relevant training in school. This sample actually included some Nobel laureates in both the sciences and literature: Niels Bohr, Karl von Frisch, Otto Hahn, Frédéric Joliot-Curie, and Ernest Orlando Lawrence in the former, and Bob Dylan, T. S. Eliot, Hermann Hesse, Jean-Paul Sartre, Isaac Bashevis Singer, Aleksandr Solzhenitsyn, and John Steinbeck in the latter.

The science/artist contrast in diversifying experiences can even be spotted in gifted adolescents. Teenagers showing scientific talent had a higher probability of growing up in stable homes in which their parents pursued conventional hobbies and interests, whereas the teens displaying artistic talent were more prone to emerge from unconventional homes that exhibited appreciable heterogeneity regarding economic mobility, geographic origins (such as foreign birth), and far-flung travels both domestic and foreign. In a sense, the family background of scientific talent is far more similar to that of the Termites, as I discussed under Tip 1. That helps us understand why the Termites included no exemplars of artistic genius.

But wait! What about the scientific revolutionaries?

Revolutionary versus Normal Scientists

When discussing Tip 2, I noted that the most eminent scientists who challenge the received paradigms tend to be more inclined to mental illness—even if not to the degree artistic geniuses are. Might not an analogous effect hold with respect to diversifying experiences? More such events and circumstances than paradigm-preserving scientists, but less than the artists? There's some tentative evidence for this conjecture. Under Tip 4, I mention the fascinating research that Frank Sulloway has conducted on whether or not a scientist will support a major scientific innovation. He examined the response to 28 innovations in all, from the Copernican revolution to the theory of plate tectonics. Some of the predictors appear close to what would be expected if diversifying experiences played a more prominent role in the development of revolutionaries—like extensive world travels, liberal social attitudes, and differences in scientific specialty. Because scientists who are receptive to innovations are also more disposed to offer their own innovations, these predictors are very suggestive, even if not positive proof.

Anecdotally, it's worth noting that Darwin lost his mother when he was 8 and Newton lost his father before he was even born. Admittedly, in the latter case it would be hard to argue that Newton directly experienced any shock. Yet the event led to his mother remarrying when he was only 3 years old, and to her dramatic decision to ship him off to his grandmother—a non-maternal act that Newton never forgave. Indeed, he once threatened to murder his mother and stepfather and burn their house down. Now that's a diversifying experience!

Closing Nature-Nurture Perplexities

At this point it would seem that the weight of the evidence favors nurture over nature—especially if we throw in the not always nurturing diversifying experiences. Surely, losing one or both parents at a young age cannot be in your genes! Nonetheless, modern behavior genetics has shown that appearances can be deceiving. What seems to be an obvious environmental effect can actually count as a clandestine genetic effect. It's almost as if the genes are lurking behind the scenes, surreptitiously exerting effects on human development—while parents, teachers, and mentors get all of the credit! To see how, I use openness to experience as a case in point—both general research findings and a very specific illustration.

Openness to Experience: General Research Findings

I already mentioned under Tip 2 that openness correlates with scores on the Creative Achievement Questionnaire, but that statement should be expanded to say that of all personality traits, it is the one that's most strongly associated with creative genius. That association becomes more comprehensible when I elaborate what it means. Openness is one of the factors in the "Big Five" personality scale, along with conscientiousness, extraversion, agreeableness, and neuroticism (to wit, the "O" part of the mnemonic acronym "OCEAN"). People who score high on openness have wide rather than narrow interests, and are imaginative, original, insightful, curious, inventive, artistic, intelligent, clever, sharp-witted, ingenious, wise, and sophisticated rather than shallow or commonplace. Although intelligence is included among the descriptors, openness only has a moderate correlation with scores on IQ tests. The personality aspects count more.

I also pointed out earlier that openness claims a substantial heritability. The genetic foundation is actually noticeably greater than any of the other Big Five, with the exception of extraversion. This heritability really throws a monkey wrench in any serious attempt to tease out the relative influence of nature and nurture. Shall we turn to a specific case study?

Openness to Experience: Very Specific Illustration

Consider William James, the American psychologist and philosopher. There's no question that he would score extremely high on openness. When young, he aspired to become an artist, even taking on an apprenticeship in the studio of William Morris Hunt, the noted American painter. Yet he soon cultivated his scientific interests by entering Harvard's Lawrence Scientific School. From there he started Harvard Medical School, but decided to take time off to join an expedition to Brazil led by the famed Swiss American biologist Louis Agazziz (after whom a building was named at Harvard's Museum of Comparative Zoology). Although illnesses prevented James from completing the trip, he decided to delay his return to medical school that he might travel to Germany, where he began to learn the latest trends in philosophy and psychology. After returning to the United States to complete his MD, he chose not to practice medicine, but instead took a job teaching physiology at Harvard. From physiology, he moved to psychology, and from psychology to philosophy, becoming a major proponent of Pragmatism. James's *Principles of Psychology* remains a classic text, a text remarkable for its breadth of coverage, including clinical as well as experimental results, comparative along with human studies. His openness is also seen in his subsequent *Varieties of Religious Experience*, a masterwork in the psychology of religion. A long-term member of the Theosophical Society, James eventually helped found the American Society for Psychical Research. If anything, many psychological scientists might find James's openness to (psychic) experiences a bit over the top!

Where did William James obtain this openness? Why not his family? From the perspective of nurture alone, it would be hard to imagine a family milieu more conducive to producing highly open offspring. William's father, Henry James Sr., was independently wealthy, living off a big inheritance, and was intellectually rather nonconformist; initial plans to enter the ministry were replaced by his becoming a theologian who advocated Swedenborgianism, a comparatively small and novel Christian denomination (the "New Church"). In fact, William's grandfather James never accepted Henry Sr.'s departure from the "true" (Presbyterian) faith. Because William's father was friends with many intellectuals of the day, such as Ralph Waldo Emerson, William and his younger siblings were exposed to a culturally rich home environment. That stimulation was augmented by the family's European tours, and his father's decision to have his children partly educated in Europe. William thereby became fluent in both French and German. This unique upbringing produced not just William James, but also his younger brother Henry James Jr., the American-British novelist, and Alice James, the American diarist. A nurture account thus sounds plausible enough.

Yet nature remains hiding behind the curtain, working in its nefarious ways! Given that openness is rather inheritable, why can't a portion of these outcomes be attributed to familial genetics? The father was obviously a very receptive soul, leaving the commonplace denomination in which he was raised to advocate another far more rare, and so why shouldn't his kids inherit the same open disposition? After all, they also seemed to acquire various psychopathological tendencies, tendencies that tend to feature substantial genetic contributions as well. Both William and Henry Jr. suffered from severe depression, the former to a degree outright suicidal, and Alice's diaries narrate in great detail her own emotional breakdowns. It's no coincidence that openness is positively correlated with cognitive disinhibition, a disposition associated with both creativity and psychopathology, as pointed out under Tip 2.

So what would have happened had William been abducted at birth and adopted out to some random family? Would he still have displayed a much higher than average openness to experience? According to research in behavior genetics, the answer is very likely, even if not quite as much. Foster children are more similar to their biological parents than to their foster parents; identical twins reared apart are more similar to each other than they are to their foster siblings. Because heritability is far from perfect, some "regression toward the mean" will occur, a phenomenon first noted by Galton. So James's openness might not be quite as extreme in this alternative universe—perhaps he wouldn't have developed any concern about psychic events! Yet the fact remains that the hypothetically adopted William James would likely exhibit a personality almost as recognizable as his physical features.

The Ultimate Message?

Do pick your spermatozoon and ovum very carefully when putting together your zygote, for roughly half of who you become may depend on that decision. Even so, exert precisely the same extreme care in choosing your home and school environments, for approximately half of your future development will hinge on that choice as well. You'll need to lean heavily on both if you want to become a creative genius. Too bad we're not given a menu at the moment of conception to make these critical selections easy. Even so-called designer babies require designed environments!

Tip 4
Be the Oldest Kid in Your Family / Make Sure You're Born Last!

OK, I admit it! I lied—or at least ended the last tip with a very misleading conclusion. Only a behavior geneticist could spot the deception. Although behavior genetics is explicitly devoted to teasing out the impact of nature, its scientific techniques are also designed to make some reasonable distinctions regarding the effects of nurture. And the discipline's most critical distinction with respect to the latter is the contrast between *shared* and *non-shared* environments. The former is what children have in common because they belong to the same family and live in the same neighborhood. William James and his brother Henry James Jr., born little more than a year apart, certainly shared the same environment in this sense: same mother and father, same household, same neighborhood—even on family trips abroad the two brothers stayed together. The non-shared environment, in contrast, is every experience that two siblings do not have in common when growing up. For instance, one sibling might be favored by the mom, another by the dad, and thus raised very differently despite having the same parents. Siblings will also interact differently with different siblings: William and Henry had exactly the same number of sibs (three brothers and a sister), except that Henry's experience of being a brother to the others was certainly unique from William's, and neither could really know what it felt like to be the brother of each other! Siblings will frequently have different friends as well, differences that become more pronounced after they enter school and find themselves interacting with contrary peer groups. In elementary school, an age difference of even one year might as well equal a decade (especially when it comes to birthday party invitations). Of course, once kids become schoolchildren, siblings can end up with diverse teachers and mentors as well—making the non-shared environment even more unshared. So the impact of nature being what it may, William and Henry might still have seen their lives diverge, one becoming an eminent thinker, the other a celebrated creative writer. Different nurture, different outcome.

Now comes my confession of deceit: when explaining the supreme openness of William James, I implied that the home environment that he shared with his siblings might deserve some credit, notwithstanding the high heritability of openness. Hence, William, as well as Henry and the other James children, obtained their openness from two sources: the genes they received at the moment of conception and the stimulating household to which they were all exposed. The problem with this twofold nature/nurture account is that behavioral geneticists have well established that the shared environment usually has little or no effect on a person's development. Certainly any similarities among sibs in openness cannot be attributed to that cause. Only specific interests and values (such as religiosity) might be ascribed to having the same family environment—and even then the effects are rather small. As a consequence, the fact that the James parents provided a very stimulating and diverse home was irrelevant. The main contribution that the parents made to their children's propensity for openness should count as nature, not nurture.

Besides, what we really seek in understanding genius is why siblings from the same environment differ so much. Not every child in the James family attained high distinction. Just compare William and Henry with their younger brothers Garth Wilkinson and Robertson. The latter are sufficiently obscure that if you google their names, you get the wrong people! For instance, googling "Garth Wilkinson James" gives you James John Garth Wilkinson, a notable Swedenborgian who so strongly influenced William James's father that he named his third son after him. And "Robertson James" just yields a list of people named James Robertson—patently the wrong persons. And even if William, Henry, and Alice attained some acclaim as writers, they did so in rather divergent ways. Why the difference? Why didn't they all become competing psychologists, or novelists, or diarists? Was nature the only cause? Or did nurture get involved via non-shared environmental effects?

A potential answer is implicit in the previous paragraph. The James children were all born in different orders: first William in 1842, then Henry Jr. in 1843, followed by Garth Wilkinson and Robertson in 1845 and 1846, respectively, and lastborn Alice in 1848. Conveniently, the first four were all sons, so differences in developmental outcomes cannot be pushed away as mere gender effects. The birth years are also very close for the siblings—five offspring in seven years!—thus maximizing any input from the shared environment. Alice's parents were not that much older when she was born than they were at the birth of her brother William. So were their distinct ordinal positions responsible?

We're now inevitably taken to one of the oldest and most controversial research topics in all of developmental psychology:

birth order. Amazingly, the story begins with Francis Galton, our already familiar pioneer in the science of genius.

Primogeniture: The Lucky Firstborn Child

The questionnaire that Galton distributed to Fellows of the Royal Society included the following item: "How many brothers and sisters had you older than yourself, and how many younger?" That's it! Not a very finely differentiated question by today's survey standards. For example, no provision was made for whether the respondent actually grew up with those brothers and sisters. Accordingly, it's difficult to know how, say, the Scottish physicist James Clerk Maxwell responded when he had no younger siblings and an older sister who had died in infancy, and thus whom he never even saw, even less interacted with. But novel research has to begin somewhere. Fortunately, researchers have advanced beyond this crude beginning.

Firstborn Sons

In his 1874 book *English Men of Science*, Galton summarized the results for the 99 scientists who provided answers: "Only sons, 22 cases; eldest sons, 26 cases; youngest sons, 15 cases. Of those who are neither eldest nor youngest, 13 come in the elder half of the family; 12 in the younger half; and 11 are exactly in the middle." He then concluded: "(1) that elder sons appear nearly twice as often as younger sons; (2) that, as regards intermediate children, the elder and younger halves of the family contribute equally; and (3) that only sons are as common as eldest sons." Thus his findings suggested some tendency toward *primogeniture*—and that's the very word he chose to head the section of his first chapter, "Antecedents," in which those findings appeared. Note that the daughters/sisters disappeared from the calculations. Hence, Maxwell simply became an "only son" whether or not he had reported his deceased elder sister. But it's hard to know how Maxwell would have been recorded had an older brother died in infancy.

Subsequent investigations also lend support to Galton's key conclusion. Take Roe's study of 64 eminent scientists introduced under Tip 3. Fully 39 were firstborn, including the 15 only children. That's already about 61% who were born first. Linus Pauling was typical: a firstborn son with two younger sisters. Furthermore, of the remaining 25 laterborns, "5 are the oldest sons, and 2 who were second-born are effectively the oldest during their childhoods since the older children died at birth and at age 2," while another laterborn had a large age gap between him and the older brother nearest him in age. Indeed, for the rest of the genuine laterborns, an average of five years separated the subject from the brother born immediately before him. Consequently, with an exception of only six scientists, or fewer than 10% of the sample, "most of those who are not firstborn are either oldest sons, or substantially younger than their next older brothers." For example, Luis Walter Alvarez, the non-Termite Nobel laureate, was born second in his family, but was still his parent's firstborn son.

The apparent primacy of oldest sons among the non-firstborns seems to imply that firstborn daughters don't even register in the family configuration! Like a leading zero in the integer "0123," big sister might as well not be there: an unadulterated cipher in describing the non-shared environment! But the Galton and Roe samples were exclusively male. What happens if we flip the comparison by looking at a sample of eminent women instead? Will a different pattern result?

Given the pernicious influence of traditional gender roles, one might guess as much. Until relatively recently, many parents were more inclined to channel limited family resources to sons rather than to daughters—to future "breadwinners" rather than to presumed "homemakers" who will be "taken care of" by a husband (presumably another family's firstborn son). In such homes, woe to the talented sister with an older brother! Sigmund Freud was his mother's firstborn child and firstborn son, at once becoming her clear favorite. He was the only child blessed with his own room (lit by a steady lamp rather than a flickering candle, no less), plus he received an allowance to buy books for feeding his curiosity (like the desire to read *Don Quixote* in the original Spanish). He even was bestowed the high honor (normally reserved by the father) of naming his lastborn brother Alexander (after "the Great" Macedonian conqueror). When his sister Anna, who was only a couple of years younger, started taking piano lessons, "Sigi" complained that her playing interfered with his studies, and the piano went out the door. He even made it known to poor Anna that he did not approve of her tastes in reading, which included works by great French novelists like Honoré de Balzac. I'm sure many little sisters throughout the world have some unprintable names for big brothers like him!

But let's move on; we're overdue for a switch in gender perspective!

Firstborn Daughters

The psychologist Ravenna Helson, who studied both eminent female mathematicians and graduates of the all-female Mills College, made a fascinating observation with respect to the Mills group: those "who were successful in careers at age 43 were, with few exceptions, those who did not have brothers." Not only were Helson's mathematicians also less likely to have brothers but also were found to strongly identify with their fathers, an identification possibly facilitated by the fact that their fathers were far more prone to be professionals. In short, absent any gender prejudices, a daughter without a brother was able to procure the advantages that normally accrue to firstborn son.

Helson's earlier results received partial endorsement in a more recent inquiry into the birth order of 112 eminent female psychologists, using 74 eminent male psychologists as a control group. The women included such luminaries as Mary Ainsworth, Mary Calkins, Anna Freud, Eleanor Gibson, Leta Hollingworth, Karen Horney, Christine Ladd-Franklin, and Maria Montessori, whereas the men included such notables as Alfred Adler, Alfred Binet, Harry Harlow, Sigmund Freud, William James, Carl Jung, Abraham Maslow, B. F. Skinner, James B. Watson, Ivan Pavlov, Carl Rogers, and Wilhelm Wundt. The contrasts between the two sexes were quite remarkable. The women not only grew up in smaller families but also were more likely to have been born earlier in the sibling configuration. Indeed, the eminent female psychologists had a higher likelihood of being an only child. Even more striking was the relation between expected birth order and the size of the family. For the men who define the control group, increasing family size tended to increase their expected birth order, which makes sense because larger sibships have more available ordinal positions. In contrast, for the women, an inverted-U curvilinear function described the relation between expected birth order and family size. The peak was at being the third-born, which occurred in families having seven children. After that peak, further expansion in the number of mouths to feed will siphon off resources so that even talented women will not have an opportunity to realize their potential if born too late in the sequence. As a side note, it's a curious fact that the psychologist in this study who could claim the highest birth order was none other than Lewis M. Terman, who was 12th out of 14 children. By comparison, Leta Hollingworth, who also conducted pioneering research on high-IQ children about the same time as Terman, was the firstborn of three children—all sisters!

There's an irony here. Although the word *primogeniture* is not strictly connected to either gender, its legal use since feudal times implied inheritance by the firstborn *son*. An example is the Salic law, which only permits succession in the French monarchy through the male line. Even so, when it comes to creative genius, primogeniture may apply with stronger force to women rather than men! A sister's talent development seems ill served by an older brother running around the house. Happily, because gender roles are gradually becoming more equal over time, this birth-order contrast should eventually disappear. After all, the eminent psychologists in the above study were born between 1802 and 1952. A lot has happened in the meantime for those born since the early 1950s. Or not. Has gender equality happened yet?

Yet, Hurray for the Laterborns!

Come on! We all know that lots of creative geniuses weren't born first in the family, nor were even firstborn sons. That was already witnessed in the James family example I presented earlier: Henry Jr. was born second, and Alice was born last, after her four brothers. Beyond doubt Henry James was as much a creative genius in fiction as William James was in nonfiction, and even if Alice's personal diary cannot compare to either *The Principles of Psychology* or *The Portrait of a Lady*, her writing certainly surpassed anything achieved by Garth Wilkinson and Robertson. Hey, her diary remains in print today! Moreover, Galton's 1874 survey results found several powerful exceptions to primogeniture. His half-cousin Charles Darwin was his parents' penultimate child, the fifth after three sisters and his older brother, and followed by one sister. The most famous of Charles's children, particularly those who were both knighted and elected FRS, were part of the laterborn half of his 10 children (albeit the picture is somewhat complicated by the fact that some died young, tragedies that Darwin later thought might be due to having married his first cousin). His firstborn son, William Erasmus, was less accomplished than the brothers born later; indeed, the son's chief claim to fame was that he became the subject for his father's "A Biographical Sketch of an Infant," a landmark contribution to child developmental psychology. Galton's ordinal position departs even more from primogeniture, for he was born "dead last" after four sisters and two brothers, all of whom survived infancy and none of whom attained anywhere near his level of acclaim. One must wonder what Galton himself thought when he discovered that neither he nor Darwin were typical (generally speaking) of FRSs.

In truth, even a firstborn son—for such I am—must admit that the presentation has so far been biased in favor of primogeniture. The scale can be tipped toward the laterborns if we weigh the following three qualifications: family size, conspicuous exceptions, and domain differences.

Family Size: How Many Sibs Do You Have?

Too often empirical inquiries confound birth order with family size, or rather sibship size. Obviously, a one-child family cannot have anything other than an only child. So why not consider an only child as both firstborn and lastborn, indicating a distinct category altogether? For example, if the 15 only children are removed from Roe's tabulation because they lacked siblings, then the percentage of firstborns with younger siblings shrinks to 38%, a far from impressive percentage. Speaking more broadly, if creative genius were randomly distributed among the ordinal positions, then 50% will be firstborns in two-child families, about 33% will be firstborns in three-child families, and so forth. Hence, firstborns can predominate if creative geniuses tend to come from small families—and they tend to do so. Admittedly, empirical studies published long after Galton and Roe have controlled for family size, and in many instances the proportions still statistically favor the firstborn, but not in all instances and not by nearly so much as so far indicated.

Conspicuous Exceptions: Any Complicating Factors?

So many exceptions exist to the primogeniture effect that a person's birth cannot become a fate-deciding event. We've already seen examples, such as the offspring of Charles Darwin. Nor should we be surprised by these exceptions. For one thing, ordinal position is a complicated factor: the effect of the actual order of birth is modulated by a host of factors, such as the temporal spacing of the births, the intrusion of half or adopted sibs, the specific configuration of brothers and sisters, and the actual longevity of each sibling, particularly whether they died before the target person could even establish a personal relationship. Creative genius is equally a supremely complex phenomenon, with multiple causes that necessarily diminish the impact of any one cause. Among those causes, naturally, is genetic endowment. And despite the fact that the non-shared environment almost always exerts more influence than the shared environment, the effect of the former remains somewhat small. If nature explains approximately 50% of who we are, that does not mean that non-shared environment accounts for the rest—far from it. And birth order is just one narrow sliver of all developmental experiences that are not shared among siblings. Early peer relations are extremely influential, too. Parents are right to worry about whether their child is hanging out with the wrong crowd! Many gifted teenagers get derailed from the development of their unique talents because of peer pressures. Having the high school reputation as the math nerd rather than the star athlete is seldom cool.

Remember J. S. Mill, who we talked about under Tip 1? The amazingly smart guy with an IQ around 200 who "had no childhood"? I left out a key biographical datum: he was homeschooled by his father, the eminent English philosopher James Mill, who wanted to nurture a genius to become a future advocate of Utilitarianism (an odd parental aspiration which actually worked). One draconian feature of that parental schooling was that his father-teacher never let his firstborn son-student interact with children his own age! No peer pressures were ever allowed! Any wonder then that the son eventually became an extremely depressed teen who pondered suicide by the time he reached 20 years old? A teenager without peers is vulnerable like a bubble kid unexposed to the biotic environment!

Domain Differences: When Do the Youngest Enjoy the Edge?

In some creative domains, the anticipated primogeniture effect flips, so that the advantage shifts to the laterborns, sometimes even to the sibling being born last. The studies discussed above all concerned the sciences, where firstborns often hold the edge. A similar pattern can be observed in classical composers as well, although exceptions always exist, such as Johann Sebastian

Bach, his father's eighth and last child. In stark contrast, a study of 64 eminent American creative writers, including notables in prose and poetry, when making a direct comparison with Roe's statistics for her 64 scientists, found a significant tendency for the writers to be born later in the ordinal sequence. Indeed, over almost three quarters of the writers were laterborns—like Henry Jr. and Alice both getting themselves begotten after William, the MD and psychological scientist in the James family. Messing up the pretty picture even more, scientists who more willingly accept major innovations—and who are also more prone to advance innovations themselves—tend to concentrate in the later birth orders. As noted earlier, both Darwin and Galton were laterborns; and because Galton was among the first to accept Darwin's theory of evolution by natural selection—congratulating his cousin less than a month after *Origin* hit the bookstores—Galton was directly inspired to contribute major Darwinian innovations of his own. Finally, just when you'd think that the empirical results cannot get more complex, sometimes *both* firstborns *and* lastborns are favored over middle children. Indeed, that curvilinear U-shaped function was even found for the Terman's gifted children: although high-IQ children were disproportionately firstborns, among those Termites that came from large families, the youngest outnumbered the middleborn children. Recall that Terman himself was the 12th born—do you have to be very bright to create an IQ test?

To make sense of these ins and outs, we really need to know how a child's ordinal position might affect their early development as a creative genius. Is there an underlying cause behind the influence? Was there a specific reason why the firstborn in Henry Sr.'s family would become a thinker and the second born a creative writer, or could it have worked just as well the other way around?

Why Would Your Birth Order Even Matter?

Psychologists love to study birth order. Almost everybody knows their birth order, and can report it almost instantly on a questionnaire. Quick: What's yours? Moreover, as personal data go, most survey respondents or experimental participants are willing to provide an honest answer to the question. In Galton's 1874 survey of Royal Society Fellows, the response rate to this birth order query approached 100%. Perhaps a few Fellows couldn't or wouldn't provide an answer because of special circumstances. Although FRS and Royal Society president Isaac Newton had passed away long before Galton conducted his investigation, it's hard to imagine how he would have answered. As revealed earlier, his mother abruptly sent him to live with her mother when she remarried, raised her children by her second husband in a separate household, and did not rejoin her firstborn son until widowed a second time years later. So was Newton then an only child? And how would a foundling—with three-dozen older and four-dozen younger surrogate sibs at the orphanage—respond to Galton's question?

Unhappily, the ease of assessing birth order far outstrips the psychologist's understanding of what underlying phenomenon is assessed by those measures. Why bother to measure it in the first place? Other than representing an unequivocal instance of a potential nurture effect, what developmental processes does it capture? Do we really need to include birth order among the nine tips? To address these issues, below I examine how a youth's ordinal position in the family might affect intelligence, personality, and opportunities.

Does Birth Order Influence Your Intelligence?

A fundamental assumption of the nurture explanation is that highly intelligent people grow up in intellectually stimulating families. Most often, this account is cast in terms of the shared environment. But is that justified? After all, siblings developing in the same home are not necessarily experiencing the same intellectual stimulation. The firstborn seems to begin life with the optimal familial milieu, for that child most often gets to monopolize the attention of two mature adults. The second-born, on the other hand, has the environment somewhat diluted by an older and immature sibling, a maturity decrement augmented for the third and later children. More, in the way of infants and toddlers, yields less. Hence, intelligence should decline with birth order. Yet this dilution of cognitive nurturance should diminish when the siblings are more widely spaced in age, even allowing older sibs to become teachers for younger sibs. But the situation can worsen in one-parent families. Twins will also be at a disadvantage, no matter whether identical or fraternal, because they expose each other to abundant "baby talk" when their parents are not looking.

What I've just done is provide a thumbnail sketch of the "confluence theory" proposed by the eminent psychologist Robert Zajonc. The sketch is oversimplified because the theory is partly expressed mathematically, but you should get the idea. Yet multiple problems confront use of this theory to account for birth-order effects in creative genius. One problem is that psychologists are still debating about whether the theory explains the development of intelligence apart from various sources of artifact. For example, laterborns may be less intelligent than firstborns, but the former must necessarily come from large families, and large families tend to be more typical of lower socioeconomic strata, confounding the child's intelligence with other factors, such as education and parental intelligence.

A more crucial problem is that confluence theory seems insufficient as an explanation even after various artifacts are controlled. To illustrate, an unusually sophisticated study of more than 200,000 Norwegians in their late teens found that the average IQ of a firstborn was just 2.3 points higher than that of the sibling born second in the same family. That difference appears minuscule in the context of the IQs estimated for creative geniuses. Henry James Jr. is definitely not going to begrudge big brother William his extra 2.3 points, a difference that falls within the plus or minus (error) of most IQ estimates anyway! Or to use another example, Albert Einstein's sister Maja (Maria) was born about two years after her older brother. They were very close throughout their lives, Albert having a very different attitude toward Maja than Sigmund Freud had toward his sister Anna, who was also about two years younger. Although Maja did not attain anywhere near the eminence of her older brother, she wasn't an academic slouch either, having done advanced studies in Romance languages and literature at universities in Berlin, Paris, and Bern, where she earned her PhD for a dissertation that was later published in a leading humanities journal. The difference between her scholarly achievement and her older brother's roughly contemporary work on the epochal General Theory of Relativity cannot possibly be ascribed to a puny 2.3-point IQ difference. Anyhow, we already noted under

Tip 1 that performance on IQ tests has a very tenuous connection with achieved eminence, our preferred criterion for creative genius.

So we've got to look elsewhere.

Does Birth Order Modify Your Personality?

I'm sorry if you're a Freud admirer, but I've got to rag on Sigmund once more. Aside from what has already been said, he was apparently a horribly obnoxious older brother. Younger sister Anna revealed as much in her "eulogy" for him after he died of cancer, remembering him as "pompous, pedantic, and priggish." I guess she never forgave him for vetoing her piano lessons! Yet Freud's treatment of his younger sibs seemed to presage his future interactions with professional colleagues. This extension became particularly conspicuous when dealing with Alfred Adler. Adler was not only younger than Freud by more than a dozen years, but he also had a rather antagonistic relation with an older brother who was not unlike Freud himself—a mother's favorite who fully exploited that status.

Nevertheless, the Freud-Adler relation went quite well at the beginning. Adler was even named the very first president of the Viennese Analytic Society. Yet soon after that promotion, their relationship started to sour, and within a year Adler resigned his position to found a rival form of psychoanalysis called Individual Psychology. Adler's departures from Freudian orthodoxy were many, but the most telling for our current purposes was Adler's advocacy of birth order as a major factor in personality development. If valid, that factor could undermine the causal primacy that Freud placed on the Oedipal complex and his psychosexual stages (oral, anal, phallic, latency, and genital). On the one hand, because the firstborn gets "dethroned" upon the birth of the second child, the former becomes hostile and insecure, eventually attempting to win back parental love and approval by becoming the most conventionally successful offspring—doing rather well in school and entering a prestigious profession as an adult: the child who pleases mom and dad by becoming a brain surgeon or rocket scientist. On the other hand, the youngest becomes spoiled, with subsequent adulthood behavioral difficulties, because of his or her excessive pampering within the family. For this reason, the youngest are supposedly inclined to become revolutionaries who seek to overthrow the tyranny of firstborn leaders. And the middle child? Can we guess? Perfection, or at least better adjusted than sibs either older or younger. It's very fortunate (even if a little suspect) that Alder himself was a middle-born child! In any case, his theory provides a personality basis for primogeniture, and even helps explain why Freud's achieved eminence is appreciably greater than his own.

Yet here's the catch. Birth order has minimal if any impact on personality development! At best, the effect is even smaller than that found for intelligence, about half as small. Surprise! Adler's theory, like Freud's, was based on qualitative impressions of single clinical cases rather than quantitative and objective research using hundreds of thousands of participants. So, again, we need to keep looking.

Does Birth Order Determine Your Opportunities?

The historian and psychologist Frank Sulloway, whom I mentioned briefly under Tip 3, would not be too pleased with the conclusion that ended the last section. In his ambitious research on the factors that lead eminent scientist to accept or reject major innovations, he has been a strong advocate of the link between personality and birth order. In particular, he has argued that birth order is positively correlated with openness to experience. As a consequence, laterborn scientists should prove more open to revolutionary ideas—and perhaps even come up with their own as well. Darwin and Galton were used to illustrate this point earlier when discussing the assets of being a laterborn. Yet if the impact of ordinal position is trivial or zip, then must we throw Sulloway's argument out the window? Not quite!

Provocative Evolutionary Theory The core of Sulloway's theory of scientific innovation is actually his Darwinian perspective on sibling development within the familial environment, a view that can be divorced entirely from any supposed connection between birth order and openness to experience or any other personality characteristic. The effects of ordinal position within a family follow a pattern of divergence that Darwin first observed on the Galápagos Islands, with their famous finches—more than a dozen species varying greatly in size, shape, and behavior, particularly with regard to food consumption. Despite their diversity, Darwin eventually inferred that all the several species had evolved from a single species that came over from the South American mainland two million years earlier, soon after the volcanic islands emerged from the sea and began to support sufficient flora. When the first colonizers arrived, they found islands replete with opportunities, varied niches that could support abundant life. Yet to fully exploit the available niches, the finches had to adapt to the diverse resources, and courtesy of natural selection those adaptations led to the origin of new species from the founding species. The upshot of that evolutionary divergence was "7 species of insectivorous tree finches, 2 species that consume the flowers and fruits of cactus, 1 species that eats fruits and leaves, and 4 species of ground finches that have their beaks graduated according to the size of the seeds they consume." Such divergence was not confined to these lowly birds, of course: the same evolutionary process yielded the awesome diversity of dinosaur species that occupied a myriad of niches in land, sea, and air before their catastrophic extinction. Galápagos finches, which underwent their own global diversion, provide a microscale illustration of a species that survived.

As Sulloway observes, "Like Darwin's Galápagos finches, human siblings tend to diversify in adaptive ways. Whereas Darwin's finches have diverged phylogenetically, through the gradual evolution of genetic differences, human siblings become increasingly dissimilar during ontogeny, through learned differences in family roles, strategies, and other behaviors." Siblings are adapting to particular family niches in order to maximize their share of parental resources, whether those be attention, praise, gifts, bedtime stories, allowances, a Band-Aid, or just a hug. In most families, the prime niche is reserved for the firstborn who will receive ample rewards, but who will also have to live up to rather demanding parental expectations, such as always respecting authority, excelling in school, and pursuing a profession. Sigmund Freud was a prototypical firstborn son, getting top grades at the gymnasium, where he graduated summa cum laude at age 17, and then going to the University of Vienna, where he earned his MD and started his medical practice shortly after. In families that don't have sons, or at least not a son born early on, the firstborn daughter may take over that niche, and thereby yield a highly achieving woman, like the eminent female psychologists discussed earlier. Nearly two-thirds of them were firstborns, counting those who were only children. That's an exceptionally high percentage.

Naturally, if mom and dad quit procreating after their first kid enters the world, the story ends. Yet as each child is added beyond the first, each addition must find a new familial niche to exploit resources not already reserved for the firstborn occupant. If your older sib is a brilliant student who's obviously headed to an Ivy League college, then perhaps you can opt for sports, becoming a star player who gets admitted to a Big Ten school on an athletic scholarship. If those two niches are already taken, then why not take up the guitar and join a rock band? Whatever the specific decisions may be, sibling competition will result in divergence not unlike that seen in Darwin's finches.

Notice, too, that this divergence can entail the attainment of roles that become not just more unconventional, but even more risky. The pathway to getting a degree in medicine, law, or engineering is well established, with many intermediate guide-posts, whereas the route to becoming a rock star is far from well delineated, and full of unforeseeable obstacles along the way. There existed no professional manual dictating that successful rock bands had to have a two-year stint playing up to eight hours per day in Hamburg strip clubs, and none that provided guidance on how to get a good drummer or what was the best hair style to replace the standard greased-up Elvis Presley look—but that's what The Beatles had to do before becoming one of the greatest rock groups of all time.

Fascinating Empirical Evidence Considerable research provides support for this interpretation. We've already gone over the tendency of great scientists to be firstborns, whereas great creative writers have a higher likelihood of being laterborns. We also noted that political revolutionaries are more likely to be laterborns, but we might as well record now that if firstborns go into politics, they're more likely to become status quo politicians, such as presidents and prime ministers in established governments. Laterborns are also more prone to compete in dangerous sports, and even when they participate in relatively safe sports, they are more disposed to engage in high-risk behaviors, such as stealing bases in professional baseball. And what about the tendency for classical composers to be firstborns? That also represents a relatively conventional and low-risk domain, especially when a firstborn can follow a fairly well-defined career path: taking lessons from the best teachers, attending the top conservatories, winning the most prestigious competitions, picking the best agents, and establishing tight connections with the right orchestras, churches, and opera houses.

Potential Overextension So far so good. Yet at this juncture Sulloway extends the family divergence principle a step further. "Such behavioral differences," he writes, "eventually become encapsulated in personality." Hence arises his claim that "firstborns are achievement oriented, conscientious, hard-working, organized, reliable, responsible, scholastically successful, and self-disciplined" whereas "laterborn[s] are attracted by novelty, liberal, prone to fantasy, nonconforming, rebellious, and unconventional." Accordingly, firstborns should score higher on *conscientiousness*, and laterborns score higher on *openness to experience*, two dimensions of the Big Five (the "C" and "O" of OCEAN). Although Sulloway provides some empirical evidence to support his view, we've already pointed out that much contemporary research argues for minimal connections between ordinal position and personality.

Still, sibling divergence can take place without stipulating any corresponding changes in personality development. That disjunction can happen because the developmental changes to siblings may very well be highly specific to the adopted domain of achievement. It must be emphasized that the Big Five personality dimensions are rather broad in conception. For instance, openness to experience entails six facets, namely, openness to actions, aesthetics, fantasy, feelings, ideas, and values. Any given creative genius might be high on some facets and low on others, depending on the domain and the creator's personal disposition. For example, the French philosopher Jean-Jacques Rousseau rates very high on openness to fantasy, feelings, and aesthetics, but noticeably lower on openness to actions and ideas, and even below average on openness to values. A value like human freedom was absolutely sacred to this great mind.

Matters can become more finely differentiated than this. As an example, even within the aesthetics facet, an artistic genius might still prove closed-minded regarding specific styles or genres. These prejudices make possible the frequent debates and controversies over new movements in the arts. An instance is the "War of the Romantics" between conservative and progressive classical musicians in the latter half of the 19th century, a controversy that pitted composers like Johannes Brahms and Clara Schumann against Franz Liszt and Richard Wagner. Among other differences, the conservatives objected to extravagant program music that seemed formless, whereas the progressives derided the stodgy adherence to outdated classical forms. The controversy could become nasty, with hisses and boos at premiere performances!

Even more finely tuned distinctions are possible. Many creative geniuses can be extremely narrow-minded about certain foods. Eat this, but not that, ever. A famous case is the American innovator Steve Jobs, who thought a strictly vegan diet insufficiently restrictive. So he substituted his own narrow diet of apples and carrots—and even engaged in outright fasting. Jobs was decidedly not the kind of guy you'd take out to dinner at an ethnic restaurant or a steakhouse. He was thus very selective in his openness.

The bottom line: birth order may help influence what opportunities future creative geniuses decide to pursue, but without leaving any huge impact on either intelligence or personality. Abilities and traits are probably much more influenced by genetic endowment and perhaps other aspects of the non-shared environment, such as peer relationships.

My Advice?

Without any doubt, the forecast provided by a person's birth order is more precise and reliable than even the best astrologer's horoscope. But it's still not good enough to be defensible in a court of law. No prospective university employer or Nobel Prize selection committee can demand that your birth order be placed right below your birthdate on your résumé or curriculum vita. So you might as well leave it off. Save that biographical fact for your future biographer who will love to narrate the sibling rivalries that helped make you who you are. And you big brothers out there: start being nicer to your kid sister!

Tip 5

Study Hard All Day and Night / Indulge Your Wide Interests, Hobbies, and Travels!

From time to time, some creative genius publishes a work that completely transforms a culture or civilization. The work becomes a pivotal point that demarcates a "before and after" spot in history. Charles Darwin's 1859 *Origin of Species* was one such landmark, and perhaps Sigmund Freud's 1900 *Interpretation of Dreams* counts as another. Western thought was dramatically different before and after the theory of evolution as well as before and after the theory of the unconscious mind. Yet the publication that launched the scientific revolution came out even earlier: *On the Revolutions of the Celestial Spheres* by the German Polish astronomer Nicolaus Copernicus. Before 1543, the earth was the center of the cosmos; after 1543, the earth was demoted to the status of a mere planet, while the sun was promoted to the center—with huge repercussions for theology and natural philosophy. The repercussions were so immense, in fact, that the Italian scientist Galileo Galilei faced the Inquisition and died under house arrest just for advocating the heliocentric system. This tragic episode much later became the basis for the famous play *Life of Galileo* by the German dramatist Bertolt Brecht. We're not talking about scientific esoterica here! Many powerful and creative people cared.

Unlike *Origin* or *Interpretation*, however, *On the Revolutions* is not an easy read, not by any means—something I can vouch for from personal experience. Indeed, it has been misleadingly branded as "the book that nobody read." Some people have read it, no doubt, but only an intellectually elite few with the requisite expertise, interest, and doggedness. Originally written in the scholarly Latin of the time, and filling up six multiple-chapter books, the masterpiece is full of sophisticated mathematics, astronomical observations, and philosophical thinking. Copernicus had clearly done his homework, showing complete competence in Ptolemy's much earlier geocentric system, plus mastering all of the pertinent astronomical contributions since antiquity, including the innovations of Islamic astronomers. For anyone with the ability to study the work, and with sufficient openness to appreciate its novel arguments, it was evident that the old system, which had dominated astronomy for more than a dozen centuries, finally got some very serious scientific competition. Yet the fact remains that not all of the 400 copies put out in the first printing were quickly sold. Contrast that non-bestselling result with the first printing of Darwin's *Origin*: its 1,250 copies sold out on the very first day! In all fairness, though, the first printing of Freud's own masterwork ventured only 600 copies and yet, despite the notion of dreams as erotic wish fulfillment, it required eight years to exhaust the inventory! Sex doesn't *always* sell.

The story is often told that Copernicus was presented with the first printed copy on his deathbed, which caused a brief respite from his terminal coma. Whether true or not, he did die the same year the book was born. More importantly, he had been working on the subject for most of his lifetime. Certainly not a spur-of-the-moment book written after a flash of inspiration during a random summer vacation! In his late teens, he started his college education at the University of Kraków (currently Jagiellonian University), which back then was the seat of a significant astronomical-mathematical school that was still in its heyday. He thereby became thoroughly knowledgeable about the latest developments. Also while at Kraków, he began to assemble an impressive library on astronomy—parts of which still survive in Sweden today. By his mid-20s, Copernicus had already made his own astronomical observations, and may even have delivered private lectures on mathematical astronomy. His studies continued until, by his early 40s, he was able to write his "Little Commentary," a short sketch of his emerging heliocentric theory. But this was only privately circulated among his contemporaries. It was not until he was a dying 70-year-old that his magnum opus saw the light of day. Consequently, Copernicus had been studying astronomy and mathematics for more than 20 years before he could so much as provide a basic outline of his revolutionary theory, and almost 30 additional years had to transpire before the definitive treatment was ready for printing. A half-century from initiation to culmination!

Sounds like tons of effort! Copernicus must have burned many gallons of midnight oil, or rather, innumerable crates of candlesticks. This arduous endeavor would also seem to offer strong support for a popular theory that creative genius only demands hard work, lots of it. Hence, we turn to the following fascinating topic.

The 10-Year (or 10,000-Hour) Rule

Back under Tip 3, Francis Galton was revealed as the first polymath in the history of psychological science. But he was definitely not the last. Born almost a century after Galton was the

American psychologist Herbert Simon, another polymath par excellence. Having earned his bachelors and doctoral degrees in political science, he expanded his contributions to encompass economics, management, sociology, cognitive psychology, artificial intelligence, statistics, and the philosophy of science. Extremely prolific, he published more than two-dozen books and nearly 1,000 journal articles. Nor were his activities those of a dilettante. He received major awards from several distinct disciplines, including the Nobel Prize in Economics, the Turing Award for computer science, the National Medal of Science, and the Distinguished Scientific Contributions Award from the American Psychological Association. When Simon passed away in 2001, obituaries appeared in the publications of disciplines ranging from public administration and economics to psychology and philosophy. Many (including me) thought him a bona fide scientific genius.

Yet from our present perspective, his most relevant scientific work began with the game of chess. Simon and a colleague started by measuring the total amount of expert information a chess player needs to advance from novice to master level. This assessment led naturally to the question of how long it takes to acquire the chess expertise necessary for world-class excellence in the game. The rough-and-ready answer was a full decade of hard practice—the so-called 10-year rule. More recently, the journalist Malcolm Gladwell announced the "10,000-hour rule," but this is just a simple bit of arithmetic. If you divide 10,000 by 10, you get 1,000, and when you divide the latter by 365, you obtain approximately 3 hours per day—a ballpark estimate of how much you need to practice each and every day. Gladwell's rule seems more precise, but it's not.

Please note that "practice" really means *deliberate* practice. For example, you don't become a chess champion by mere recreational playing or going up against an ordinary commercial computer program. You have to study books on strategies in the main phases of a chess match as well as scrutinize representative past games, including those by chess greats. When Gary Kasparov, the world chess champion, lost the momentous 1997 match to Deep Blue, the chess program, he complained that he was not allowed to study Deep Blue's prior games, whereas the computer programmers could examine all of Kasparov's past games. Unfair!

Although research on expertise and practice began with chess, Simon's colleagues and students soon extended the approach to other domains of high achievement—like sports and music performance. In time, even creative genius was viewed as the mere upshot of deliberate practice. Just review your lecture notes, highlight key terms in your textbook, do your problem sets, prepare for every exam, take your writing assignments very seriously, and you'll soon be on your way to greatness. Recall how much time and effort Copernicus seems to have devoted to mathematical astronomy?

The first empirical test of this point of view was conducted by John R. Hayes, one of Simon's colleagues at Carnegie Mellon's psychology department. The particular focus was classical composers, a group that often includes child prodigies such as Wolfgang Amadeus Mozart. Hayes began with a sample of 76 great composers, and then collected data on the age at which they began their deliberate practice in music. In addition, the investigator determined the age at which they created their first big hit—as judged by recording frequencies. For almost all, the gap between those two ages was 10 years or more. There were only three exceptions, even if minor ones: Niccolò Paganini's *Caprices*, Erik Satie's *Trois Gymnopédies*, and Dmitri Shostakovich's Symphony no. 1 in F Minor, op. 10. The masterpieces of Paganini and Shostakovich came out only one year earlier than the rule would demand, and Satie's masterwork appeared only two years earlier. Those three departures seem too tiny to cause much concern.

Yet what about a child prodigy like Mozart? Hayes demonstrated that "while 12 percent of Mozart's works were written in the first 10 years of his career, only 4.8 percent of the recordings came from the early period." But Mozart only lived 35 years! Even worse, these figures overstate the impact of those early works: many of these are found in exhaustive collections, like compilations of Mozart's entire symphonic output. That inclusiveness is misleading because it's certain from first listening that his Symphony no. 1 (K. 16) would not suffice to secure him immortality. Composed when just 8 years old, the work betrays obvious influences from much older composers, such as Johann Christian Bach, whom Mozart had newly met in London. When Hayes deleted complete recordings from his tabulations, and stipulated that a composition had to feature five distinct recordings to count as a masterwork, Mozart did not produce anything of note until he was 12 years into his career. The 10-year rule seems vindicated accordingly.

Anders Ericsson, one of Simon's most successful students, has recently argued that that deliberate practice is essential for all peak performance, the supreme achievements of the creative genius not excepted. Composing a great symphony or opera is not that different than winning a chess championship. Let's see if that's really true!

Rules Are Made to Be Broken

To be honest, I've had face-to-face conversations with both Simon and Ericsson about the issues treated here. Both teacher and student see no scientific value in research on individual differences other than those that result directly from nurture. And deliberate practice can be considered an act of self-nurturing—it requires rigorous self-discipline, that's all. If you're not already a creative genius, you're just lazy, and who knows where laziness comes from. At a conference breakfast in 1995 Simon instructed me in all seriousness that the findings of behavioral genetics just could not be believed. And at a 2011 dinner reception before a joint public debate at Pomona College, Ericsson seriously questioned the validity of individual difference measures, such as intelligence tests.

Yet these strong stances are not idiosyncratic to these two illustrious psychologists. In 1957, one year after Lewis M. Terman left the world, one of his Termites, Lee Cronbach (I mentioned him in passing under Tip 1), published "The Two Disciplines of Psychology." In that article he emphatically distinguishes experimental psychology from correlational psychology. Where the experimental discipline dates back to Wilhelm Wundt, its founder, in the late 19th century, the correlational psychology originates with Francis Galton, who initiated the measurement of how people differ from one another. Psychologists in the two disciplines seldom speak to each other, and even less understand each other when they do speak. I know. I left that breakfast conversation with Simon amazed, and got up from the dinner exchange with Ericsson frustrated. Although I've published laboratory experiments, my psychological center of gravity falls squarely in the correlational camp.

Hence arises this "trigger warning." If you think that all persons are born absolutely equal, with no variation in intellect, character, or talent, then you had better stop reading right now. Just skip directly to Tip 6. I can't make any impression whatsoever on those who have already convinced themselves that history would not have changed one iota had the newborns Pablo Picasso and Albert Einstein somehow been switched in their cribs. But if you are open to the possibility that correlational psychology has something to offer in understanding creative genius, then read on!

Prodigies: Precocious Expertise

For experimental psychologists, individual differences are an annoyance. Participants in lab experiments do not respond identical ways to identical manipulations or treatments, but that variation is just swept under the rug as putative "error variance." But for correlational psychologists, that variation enjoys inherent interest, and the researcher will therefore try to find the "correlates" of those differences. So let's go back to the study that Hayes conducted of musical genius. Although he observes that some classical composers depart from the 10-year rule, he has no interest in explaining why, perhaps because the differences seem so small. Satie marks the low end, at 8 years, while Mozart marks the high end at 12 years (albeit these numbers are not strictly comparable, because of methodological differences). The only real interest for Hayes was that Mozart was not that much of a prodigy if it took him so long to produce a masterpiece.

Also problematic is that Hayes never informs us how many of the 76 composers took far more than a decade. After all, if creative geniuses can depart from the 10-year rule, it's important to assess those departures both ways. Musical talent can be either positive or negative. The positive talents require less than a decade; the negative talents require more than a decade. Indeed, in extreme cases of negative talent—the complete lack of talent, that is—a person might take a lifetime trying to attain mastery and yet never make it big time. These poor souls may not enter the history books, but their number might actually be very large. I know a few personally, sad to say.

The 120–Classical Composers Study In light of the foregoing problems, I conducted a new and improved study a few years later. The sample of classical composers was expanded to 120, which means that the sample included more obscure contributors to the repertoire. I hate to name names, but I will anyway: like Josef Suk, the Czech composer who was the former pupil and son-in-law of the far more famous Antonín Dvořák. The range of creative talent would thus be much wider than in Hayes. The sample was also historically broad, going from Giovanni Pierluigi da Palestrina (born circa 1525) to Benjamin Britten (died in 1976), more than four centuries of music history. Collectively, the 120 musical geniuses were responsible for almost 90% of all compositions heard in the classical repertoire. That's darn near representative of everybody whatever his or her genius may be. If you're a classical music buff, can you even *name* 120 composers off the top of your head?

Unlike in the earlier study, the onset of "deliberate practice" was assessed two different ways: the age when music lessons began and the age when composition started. The latter could include those juvenilia that often embarrass the composer after entering maturity (or that the composer may commit to the flames to prevent future embarrassment, as Johannes Brahms did). These two assessments were then compared with a host of career variables, including the age of first hit, the maximum annual output, total lifetime output, and achieved eminence. By "hit" and "output" I mean works that actually entered the standard repertoire rather than works that fell by the wayside. This criterion remains more inclusive than the one Hayes used, because he required that the composition be considered an authentic masterwork with at least five recordings. Yet there are many fine pieces that have fewer than five in-print recordings in the major catalogs, particularly if one early recording was widely recognized as the definitive performance.

Significantly, the variation in the onset of either first lessons or first compositions is huge. Lessons could begin anywhere between age 2 and age 19, and composition could start anywhere between age 4 and age 32. The study's most remarkable 2-year-old was the French composer Camille Saint-Saëns, who began piano lessons with an aunt who quickly recognized his precocity. The same prodigy also started composing at age 4, producing a small piano piece with the autograph score enshrined in the Bibliothèque nationale de France.

To test the 10-year rule, we must combine these two developmental variables with the composer's age at first hit—the very first work to enter the standard repertoire. We just subtract the age at first lessons from age at first hit to get a measure of *musical preparation*, and subtract the age at first composition from age at first hit to get a measure of *compositional preparation*. The statistics for both new preparation measures are very revealing. For the first, the scores ranged from 2 to 43 years, with an average of about 19 years. Not only is the variation immense, but also the average is almost double what's expected by the 10-year rule. For the second measure, the values range from 0 to 41 years, with a mean of about 12 years, which better fits the rule. But, still, the range is enormous—almost two thirds the average life span for the 120 creators.

Naturally, the composers with zero years of compositional preparation were not the same as those with zero years of musical preparation. In fact, the former were most likely musicians who had taken lessons for many years, and even had commenced careers as concert pianists or violinists, before trying their hand at composing. They might then have used very familiar works, such as encore pieces, as models, and within a year managed to produce an enduring work even if not a masterpiece. Hayes said that Paganini wrote his 24 violin *Caprices* only nine years into his virtuosic career, but that career consisted almost exclusively of playing the violin, sans formal compositional training. The *Caprices*, all 24 of them grouped and numbered, are identified as Opus 1.

In any case, the 10-year rule is already broken. But that break shatters into minuscule fragments when we calculate the correlations between these two preparation measures and three critical indicators of creative performance. To be specific, those composers who require *less* time to acquire the necessary expertise—whether in musical or compositional preparation—tend to reach a *faster* maximum output rate, accumulate *more* total hits, and attain a *higher* level of eminence. In short, less equals more. In terms of the 10-year rule, the less time consumed in deliberate practice, the higher the likelihood of becoming a creative genius.

Talent can be defined as an innate capacity to accelerate the acquisition of domain-specific expertise. That definition parallels the old concept of IQ, where a high IQ means that the mental age is older than the chronological age. The greatest classical composers thus must boast high "compositional IQs" that speed up both musical and compositional preparation. They're just better faster.

One Classical Composer Exemplar Felix Mendelssohn may offer a typical case. At 6 he began taking piano lessons, and by 10 he started studying compositional techniques. Between ages 12 and 14 he wrote a dozen string symphonies that are frequently broadcast on classical music stations and have been recorded multiple times—*nine* in-print recordings according to a recent online check. So the first symphony was written with six years of musical preparation and two years of compositional preparation. That should be indicative of precocious expertise. To be sure, Hayes might argue that the string symphonies are not true masterpieces. For that requirement to be met we have to wait until the String Octet in E-flat Major. At that time Mendelssohn was 16 years old, a decade after starting piano lessons and six years after he began learning compositional methods. That still sounds extraordinary to me.

Moreover, after his compositional career took off, he created at a prolific rate, producing a large number of masterpieces despite only living until he was 38 years old. Besides the Octet, his "warhorses" alone include the overture and incidental music to Shakespeare's *Midsummer Night's Dream* (containing the ubiquitous "wedding march"), the "Scottish" and "Italian" symphonies, the *Hebrides Overture* ("Fingal's Cave"), the Violin Concerto in E Minor (which has been ranked right up there with concertos of Beethoven, Brahms, and Bruch), the Piano Trio no. 1 in D Minor (his greatest chamber work besides the Octet), and his famous *Songs Without Words* written for solo piano. A boxed set called "Mendelssohn: The Masterworks" holds 40 CDs, or an average of about one CD of hits for each year of his life! That's more than today's pop composers.

Lastly, even if Mendelssohn did not reach the distinction of Bach, Beethoven, or Brahms, a recent study placed him roughly in the upper 3% of 523 significant classical composers—right between Giuseppe Verdi and Carl Maria von Weber. So 97% of his competitors for fame would love to change places with him! So much for the 10-year rule!

Polymaths: Versatile Expertise

I stressed under Tip 4 the crucial connection between openness to experience and creative genius. Of all personality traits, openness is the one that correlates highest with actual creative achievement. Higher scorers have broad, even artistic interests, even when those interests might appear irrelevant to the creator's domain of expertise. For example, Nobel laureates in the sciences tend to have more artistic hobbies in comparison to their less distinguished colleagues, such as non-laureate members of the US National Academy of Sciences. Sometimes these extraneous interests, artistic or otherwise, will spill over into their scientific work. According to the official Nobel website, Murray Gell-Mann's "interests extend to historical linguistics, archeology, natural history, the psychology of creative thinking, and other subjects connected with biological and cultural evolution and with learning." He attained acclaim, and a Nobel Prize, for two well-known contributions to quantum theory: the quark and the Eightfold Way. Gell-Mann spelled *quark* (a term he'd already determined the pronunciation for) based on a passage in *Finnegan's Wake* by James Joyce; the Eightfold Way alludes to the "Noble Eightfold Path" of Buddhism.

Admittedly, these two instances might seem trivial. The concepts would have remained the same had Gell-Mann used other, more abstract but still catchy alternatives, such as "subsubs" for the quarks (i.e., "sub-subatomic particles"). Yet often such wide interests make things happen that might not have otherwise—like the discovery of mountains on the moon.

The Utility of Broad Interests Galileo Galilei complemented his scientific endeavors with strong interests in both literature and the visual arts. His literary activities probably helped make his publications persuasive exemplars of Italian polemic prose, yet his artistic activities had an even more profound impact. He was not the only astronomer of his day who thought that the newfangled telescope might be usefully pointed toward the heavens. Yet he noticed things that his contemporaries overlooked. Hence, where others would look at the magnified moon and just see a smooth surface with disordered dark and light areas, like a marble, Galileo saw mountains! Not just that, but he drew pictures of those mountains that vividly demonstrated that the light and dark areas just represented peaks and valleys. How did he do that?

Galileo's artistic interests included training in chiaroscuro, the very technique used by contemporary artists to suggest relief in two-dimensional paintings. So effective were Galileo's drawings that a Florentine painter, who was a friend of his, soon incorporated Galileo's representation of the lunar mountains into a painting of the Madonna posed above the moon. If Galileo had no interest in the visual arts, it is doubtful that he would have noticed something that helped accelerate the scientific revolution.

I must confess that we've seen nothing so far that truly breaks the 10-year rule, which only amounts to about three hours per day distributed over one decade. So even after putting in a full day's work, ample "down time" should remain to indulge in an avocation or two, a timeframe devoted to talent development. And what would be the implications if a creator had not just an extra interest, but also an additional full-fledged expertise? What about those creative geniuses who achieved eminence in more than one domain of creativity? If we talk about two different domains, would 20 years of deliberate practice be necessary? If three domains, then 30 years? If so, are polymaths even possible? By definition polymaths attain distinction in multiple domains, and multiples of 10 yield numbers that can easily exceed the length of a human productive adulthood, leaving no time for eating or sleeping. How is that doable? How can polymaths like Francis Galton or Herbert Simon even exist? To answer, it is first necessary to distinguish different degrees of versatility.

Degrees of Versatility All polymaths are versatile, but not all versatile geniuses are polymaths. A true polymath is a person whose creative achievements span two or more separate domains, domains so different that they would normally seem to represent a distinct expertise. The medieval abbess Hildegard of Bingen is best known today as a composer whose music is still performed and recorded. Yet she also made contributions to drama, poetry, philosophy, theology, natural history, and medicine. For example, she wrote what is now viewed as the oldest extant morality play. Even more astonishing, Hildegard was a mystic and religious leader of sufficient importance to become canonized by the Roman Catholic Church. Other creative geniuses whose versatility extends to polymath levels are Aristotle, Al-Kindi, Abhinavagupta, Shen Kuo, Omar Khayyám, Leonardo da Vinci, Blaise Pascal, Mikhail Lomonosov, Thomas Jefferson, and Thomas Young. Such people can happen!

At a somewhat lower level, but still remarkable, are those creative geniuses whose versatility extends to two or more subdomains within a given domain. This degree of versatility is common in literature where a writer can produce masterpieces in more than one specialized genre. William Shakespeare created both great plays and poetry, for example. Of course, a writer might attempt more than one literary form, but also fail in that attempt. Thomas Hardy wrote first-rate English novels and poetry, but his ambitious effort at drama, the three-part *The Dynasts*, was not considered a masterwork either in his own time or in ours. Similar examples happen in other creative domains. The Austrian composer Franz Schubert created great songs, piano sonatas, chamber music, and symphonies, but his overly numerous attempts to put something successful in the opera houses proved in vain; these days only a few of his operatic overtures show up in the concert hall. Nor are all of the failures restricted to moves from smaller to larger forms, such as concert music to opera. It can work the other direction, too. Although the Irish author James Joyce wrote celebrated novels and short stories, his attempts at lyric poetry were not nearly so fruitful, and thus his poetry is most often omitted from anthologies of great poems.

To get an idea of the relative rarity of versatility in any form, we can turn to an empirical study of 2,102 creative geniuses in Western civilization. Fully 61% did not contribute beyond a single subdomain of a larger domain. These folks are not to be belittled, because they still produced works of genius, such as the poems of Emily Dickinson, the paintings of Vincent van Gogh, the operas of Giacomo Puccini, and the theoretical models of Albert Einstein. Another 15% contributed to more than one subdomain, while still confining their creativity to a single domain. T. S. Eliot, the Nobel laureate in literature, wrote highly regarded poetry, plays, and literary criticism, but offered no great paintings, operas, or theoretical models.

Surprisingly, the third group, the polymaths, constituted 24% of the sample. These creators contributed to more than one domain, and often to more than one subdomain within a domain. The German polymath Johann Wolfgang von Goethe not only wrote great novels, dramas, and both lyrical and epic poetry, but also made notable contributions to both botany and anatomy. When once asked what he considered his greatest work, rather than say *Faust*, which would seem the obvious answer today, he mentioned his *Theory of Colors*—an attack on Newton's color theory that had more impact on philosophers, artists, and psychologists than on physicists. Plus, Goethe accomplished all this while maintaining his time-consuming day job as diplomat and civil servant. How did he manage to acquire enough expertise to do all this? The 10-year rule just seems woefully inadequate as an explanation.

The same explanatory inadequacy holds for the other 482 highly versatile geniuses in this sample, as well as for the 302 who made their claims to fame in more than one subdomain. Some explication is especially necessary given that these creative geniuses are not mere dilettantes. Not only can versatile creators be more productive than non-versatile creators, but they can also attain higher degrees of achieved eminence. Goethe is the greatest in German literature, period.

So How Are Polymaths Even Possible? One potential answer may be derived from an idea that emerged earlier: the acceleration of expertise that permits the young talent to develop faster. If someone, let's say Goethe, is born with multiple positive talents, he (or any other polymath) just needs less time to get up to speed in multiple domains. Beyond this, the polymaths may also "get more bang for the buck" in the sense that they can achieve more with a lesser amount of deliberate practice. The personality trait *openness to experience*, which we've explored several times so far, may be one reason for this increased efficiency. Remember that greater openness is associated with more cognitive disinhibition, thus enabling the creator to notice things that others ignore—like Fleming's serendipitous discovery of penicillin, discussed under Tip 2. This trait might give an edge.

A final possible explanation for polymaths harks back to Herbert Simon. Once I asked a student of his how Simon could subscribe to the 10-year rule when his own creative career seemed to contradict it. The response? Simon proposed a single powerful principle, which he coined "bounded rationality," with multidisciplinary applications. So he could just apply that principle to economics, political science, cognitive psychology, computer science, and the rest, thereby attaining high honors in several domains. I didn't buy into that explanation back then, and I still don't. The problem is twofold. First, even if he just extended the same big idea to multiple domains, he still had to master those domains sufficiently well to convince those already trained in them to see the value in the idea. Simon couldn't just say, "Hey economists! Here's a psychological concept that deserves a Nobel Prize in your own field!" Second, I've read several of Simon's publications that have absolutely nothing to do with "bounded rationality" or any related topic. Moreover, these publications appeared in highly technical journals that demand a specialized expertise. Hence, Simon's own creative career breaks the 10-year rule. I think he was tremendously gifted, acquiring expertise faster and doing more with the expertise he had already acquired. Both better faster and more bang for the buck!

Revolutionaries: Original Expertise

The 10-year rule makes a huge assumption when it states that domain-specific expertise takes a full decade to acquire. That stipulation certainly holds for chess, a game that was played for more than a millennium before it became the basis of Simon's study. In addition, keyboard and string instruments have been used for hundreds of years, and most sports on which research has been based have been around for almost as long, such as golf and tennis. The available knowledge and techniques that have accumulated over the intervening years are immense. Mastering that knowledge and becoming proficient in those techniques requires years upon years. In music performance, for example, lesson plans become so well established that they are graded according to acquired proficiency: beginners, intermediate, and advanced, with finer gradations between. You can't do *this* before you first learn how to do *that*. Like arithmetic has to come before algebra, and algebra before calculus. And progress within these three must be broken down to more detailed steps, such as differential calculus before integral calculus.

But what happens if the domain is brand new? What if almost no cumulative knowledge exists, and techniques remain in their infancy at best? Must a creative genius really have to wait a decade before making major contributions to the new domain? Of course not! If there's nothing to master, then just go to it! To see why I arrive at that conclusion, let's look at two examples, both involving entirely antithetical forms of "looking"—one far, far away and the other really, really close up!

Telescopic Astronomy When Galileo began using his handcrafted telescope to make his epochal (and astronomical!) discoveries, telescopic stargazing devices simply didn't exist. From the ancient Babylonians and Ptolemy to Copernicus and Tycho Brahe, all observations of the heavenly bodies were conducted using the naked eye, albeit with increasingly improved instruments to make measurements more precise. Telescopes did not appear until 1609, and it was only then that they might be pointed toward the night sky. Galileo first learned of the new invention in the same year, but soon realized that the existing telescopes were inadequate for anything besides spotting enemy ships on the horizon. Because those early telescopes required two lenses, and given that the optics at that time only dealt with one lens, Galileo had to engage in arduous trial and error before he devised an instrument suitable for observing the stars, planets, moon, and sun. In fact, he didn't truly know how his device worked (a deficiency later remedied by the German astronomer Johannes Kepler). Worse still, no current domain-specific expertise then existed to guide him on what he might expect to see. Indeed, whatever expertise was available at the time was plain wrong. According to the prevailing Aristotelian cosmology, the moon, sun, and planets were all crystalline spheres with perfectly smooth surfaces, and the stars behind these moving objects were utterly fixed, providing a mere backdrop, like the "stars" on a movie set for an outdoor night scene.

Fortunately, Galileo had much earlier conducted experiments in physics that showed that the great Aristotle didn't always know what he was talking about. Heavy objects did *not* fall faster than light objects. The revered expert's expertise was subject to revision over time. Better yet, Galileo's broad interests in literary and artistic endeavors show that he must have been extremely open to experience, so he would naturally go by what he actually saw through his telescope rather than yield to what he was supposed to see according to the authorities.

Hence, as I already noted, he discerned the lunar mountains that weren't supposed to exist in the first place. When he then turned the telescope toward Jupiter, he spotted four stars that were by no means fixed, and eventually figured out that they were genuine moons orbiting around that planet, something unheard of in either Ptolemy or Aristotle. From there he went on to observe the phases of Venus, which changed in a manner most consistent with the Copernican heliocentric system. Soon he also added sunspots to his observations, yet another proof that the heavens were imperfect. In just three years, a domain expertise more than a millennium old had basically ceased to exist. Subsequent astronomy would have to start from scratch, leaving the accumulated achievements of antiquity in the trash-can. Nobody but a historian of science would ever want to pull off the shelf and dust off a volume of Ptolemy's *Almagest* or Aristotle's *On the Heavens*.

Microscopic Biology The second example is perhaps even more amazing. Although the magnifying powers of single lenses had been known for a very long time, particularly with the advent of eyeglasses in the late Middle Ages, the compound microscope that uses two lenses did not emerge until roughly the same time as the telescope. In fact, Galileo himself had toyed around with one before quickly losing interest. When the compound microscope was pointed at small organisms, the results were amazing, as seen in the wonderful plates published in Robert Hooke's 1665 *Micrographia*. These plates showed magnified images of lice, fleas, flies, and other small objects. But there was a catch: the objects portrayed were already known to exist; the amazement just resided in the minute details that couldn't be seen with the naked eye. The compound microscope was too low-powered to make discoveries that would overthrow Aristotelian biology like the telescope did with Aristotelian cosmology and Ptolemaic astronomy. That limitation was soon to change, and in the most surprising way.

Antonie van Leeuwenhoek, a Dutch draper with no training other than as a bookkeeper's apprentice, had devised a new single-lens microscope more powerful than any microscope then in existence. His original purpose was product quality control—to ensure that his textiles were made of the finest thread. Learning to make his own lenses, he invented new procedures to produce microscopes that could magnify objects 300 times, about 10 times more powerful than any compound microscope back then. His curiosity soon took over once he realized that he could view things that could *not* be seen with the naked eye, such as bacteria, protozoa, and spermatozoa; the cell's vacuole; muscle fibers; blood cells and blood flow in the capillaries; and other microscopic objects that Aristotle never knew existed. Imagine the openness to experience necessary to scrape the plaque off the teeth of two old men who had never cleaned their teeth in their entire lives! Yet that's how he discovered bacteria.

In 1673 Leeuwenhoek began communicating his findings to the Royal Society of London in his dialectical Dutch; the society secretary (who learned Dutch to do so) then translated it. But by 1676, his credibility came under suspicion when society members read his first reports of "animalcules" (microscopic life). That claim ran completely afoul of current domain-specific expertise. It didn't help that that none of these expert scientists could figure out how he made his lenses, a secret that was not revealed until more than 200 years later. So the society ended up sending a special commission consisting of both scientists and clergy to see firsthand whether his discoveries were valid and truthful. The members finally acknowledged his contributions a year later, and elected him Fellow a little after. Just as Galileo had invented the new domain of telescopic astronomy, Leeuwenhoek had invented the new domain of microscopic biology.

Revolutions in Established Styles and Paradigms Once a domain of creativity is fully established, one might think there's no longer any need for original expertise. Any scientific or artistic contribution can just become an incremental addition to received knowledge or techniques. Chess champions may devise novel tactics, but they don't change the rules of the game, and any innovations are often founded on the skills of past champions. But when a domain of creativity undergoes a revolutionary change—whether in artistic styles or scientific paradigms—an appreciable part of that traditional expertise is dismantled to make room for a new expertise. When the spirituality of medieval paintings was replaced by the naturalism of the Renaissance, the technique of applying gold leaf halos was no longer an indispensable part of the artist's tool kit. Halos became so naturalistic looking that they conformed to the linear perspective organizing the entire painting, an organizational technique utterly unknown in the Middle Ages. Of course, linear perspective itself got thrown out of many painters' kits when modern art burst upon the world in the early 20th century. Where's the linear perspective in abstract expressionism, for instance?

Go back to Kuhn's theory of scientific revolutions treated under Tip 3. Here's how he described the revolutionaries who replaced the old paradigm with a new one: "Almost always the men who achieve these fundamental inventions of a new paradigm have been either very young or very new to the field whose paradigm they change," for "obviously these are the men who, being little committed by prior practice to the traditional rules of normal science, are particularly likely to see that these rules no longer define a playable game and to conceive another set that can replace them." Although some aspects of this description will be qualified when we discuss subsequent tips, its main point remains correct. Some part of the traditional expertise must be replaced by an original expertise. Persons "new to the field"—those who did not necessarily engage in the drudgery demanded by the 10-year rule—may be the ones to carry out that replacement. Revolutionaries break the rules in creativity just like they do in politics.

Copernicus Redux

On first glance, Kuhn's sketch of the scientific revolutionary would not seem to apply to Copernicus, the guy whose creative career instigated this entire discussion about the 10-year rule. Even the surviving portraits might make him look a bit too stodgy to initiate a scientific revolution. But closer examination reveals that he might have been exactly the right genius for that epochal moment in history.

To start off, he was the lastborn child of four children, with an older brother and two older sisters. So we already know from Tip 4 that he might have some tendency, however modest, to be open to innovative ideas, if not become an innovator himself. The brother and one sister entered the church, the former as canon and the latter eventually as a prioress at a convent, but the other sister married someone active in both business and city government in their hometown. Copernicus assumed some responsibility for their five children for the rest of his life. His involvement as Uncle Nick was probably enhanced by the fact that he never married, nor seems to have fathered any natural children (he may or may not have been ordained). Even so, later in life he had a live-in housekeeper whose presence raised a big scandal—whether justified or not—and after repeated admonitions from bishops he ended the relation after nearly a decade, for whatever reason. In any case, Copernicus was obviously not one to readily conform to societal conventions, astronomical or otherwise.

Another piece of the puzzle: Copernicus was orphaned, and thus endured one of the most prominent manifestations of diversifying experiences discussed under Tip 3. More specifically, his father died when he was about 10, and shortly after he acquired a new father, namely, his maternal uncle, who made sure that his young nephew got the preparation needed to enter the University of Kraków, the uncle's own alma mater in Poland's capital.

Now that I've mentioned one diversifying experience, I must also mention another: Copernicus had a preeminently multicultural background. I hinted at that fact earlier when I called him a "German Polish" astronomer (I could just as well have said "Polish German"). He grew up in a region that was polyglot, with a strong presence of both German and Polish cultures. Although Copernicus likely spoke German in the home, he could also speak Polish—as well as Latin, Greek, and Italian—albeit not all with equal fluency. He certainly would have had plenty of opportunity to practice Italian, a language very different than either German or Polish. After finishing his studies at Kraków, he continued advanced work at the Italian universities of Bologna, Padua, and Ferrara. It was in Italy where he obtained his doctorate in canon law (not in mathematics or astronomy). A few years later he returned to Italy, this time Rome, for an apprenticeship with the Roman Curia to prepare him for getting a day job within the Church back in Poland. Not only had he lived abroad but he also was well traveled abroad. Almost a thousand miles of rugged unpaved roads then separated Kraków from Rome, with lots of mandatory stops along the way.

Italy at that time was a center of humanistic studies, so Copernicus was not remiss in expanding his expertise beyond mathematics, astronomy, and canon law. He took courses from distinguished humanist professors at Bologna. Copernicus even became a legitimate classics scholar, publishing his own translations of Greek literature into Latin and writing an epigram in Greek. Moreover, when he realized that medicine would become part of his future responsibilities, he also studied that subject at Padua, then one of the top-flight medical universities in Europe. He later became a practicing physician for his uncle-bishop. By now it should be obvious that Copernicus was a polymath. But if you're not yet convinced, Copernicus put forward two fundamental concepts still discussed in modern economics, namely, the quantity theory of money and the principle most often called Gresham's law, but sometimes named the Copernicus law instead. *He* definitely would have deserved the Nobel Prize for Economics if it were bestowed way back then.

Copernicus was similar to Goethe in that his official employment was often demanding, not a mere sinecure that would enable him to hide away night and day in his study. Copernicus on multiple occasions had to serve as politician, diplomat, and administrator. And just as Goethe still found the time to finish his *Faust*, Copernicus made room for completing his *On the Revolutions of the Celestial Spheres*. Both long-term projects were even completed in the very last year of the authors' lives. That both polymaths were able to accomplish so much in so little time delivers the ultimate coup de grace for the 10-year rule—at least if you aspire to become a polymath. If not, then just start studying, very hard and very narrowly focused! You'll then avoid becoming a dilettante, but still fall far short of creative genius in any domain.

Tip 6

Impeccable Perfectionists Rule / More Failures Mean More Successes!

One common question asked during job interviews has become proverbial: *What's your greatest weakness?* This probe usually follows on the heels of another enquiry: *What's your greatest strength?* Such an incongruous juxtaposition might seem a little disarming, except that most job applicants these days know the script and prepare their responses well in advance. One ubiquitous answer—*Well, I should admit that I'm a perfectionist*—sounds commendable enough, except when the interviewer has heard the same cliché from almost everyone else. But for those of us who have worked under or with perfectionists, we know that perfectionism is not always the desirable trait it's cracked up to be. But surely, perfectionism has a pronounced upside for the creative genius. Or does it?

To find out, I decided to google "perfectionism," which quite quickly landed me at the Wikipedia article on "Perfectionism (psychology)"—a concept distinguishable, evidently, from "Perfectionism (philosophy)." I know, I know: Wikipedia itself seldom exemplifies perfection, but it's a reasonable place to go to satisfy an urgent curiosity. Anyway, the crowd-sourced entry wisely distinguishes between negative and positive perfectionism. In the former category are such maladaptive consequences as anorexia nervosa and suicide, whereas in the latter category is the adaptive capacity for truly notable achievement. To give examples of great perfectionist achievers, the anonymous editors name the scientists Nicolaus Copernicus and Isaac Newton, the artists Filippo Brunelleschi, Leonardo da Vinci, and Michelangelo, the writers Gustave Flaubert and Franz Kafka, the composers Ludwig van Beethoven, Johannes Brahms, and Brian Wilson, the film directors Stanley Kubrick and Andrei Tarkovsky, and the entrepreneurs Steve Jobs and Martha Stewart.

No doubt various fans will later edit this collection of names to include their own favorite exemplars of perfectionism, so check the Wiki entry for later additions. Yet the main point remains: almost all (even if not all) of those identified have some claim to creative genius, which alone might imply that perfectionism supports exceptional creativity. Even so, a very large number of creative geniuses were *not* identified, and they might actually provide counterexamples. Remember Alexander Fleming, whom I discussed briefly under Tips 2 and 5? I didn't mention earlier that Fleming was less than meticulous in maintaining a tidy laboratory. This lack of care then facilitated his discovery of penicillin. Perfectionists among his colleagues would not have left an unattended stack of staphylococci cultures exposed on a lab bench before taking off on a family vacation! Still, how typical can that be? So let's look at what we know.

Why Be Perfect?

The most obvious proofs that perfectionism is essential to creative genius are found in the very products of that genius. Each singular achievement might be universally admired as a masterwork or masterpiece, a tour de force or chef d'oeuvre, or a showpiece, classic, or magnum opus. Each celebrated work would be replete with ideas, details, discoveries, or flourishes that mark flashes of genius. To illustrate, let us turn to the following two examples spanning from science to art.

• Isaac Newton was a notorious perfectionist. When two of his major publications were found to contain mistakes—one due to printer's errors and the other because the publisher used the wrong manuscript—Newton commanded that his name be utterly removed from the title pages! Newton's supreme tour de force, *Mathematical Principles*, is crammed with mathematical deductions and arithmetic calculations. Surely, there's an error or two somewhere that he failed to catch. And there was—one! About three centuries after publication, a physics major finally discovered during a course assignment that the great Isaac had inadvertently inserted the wrong number into an equation. The math was fine, but the answer couldn't come out right without the correct number. But I think we can still consider Newton a perfectionist. Perfectionism isn't always perfect.

• Michelangelo's perfectionism frequently slowed down his creative productivity because he refused to call a work finished until it had completely satisfied his extremely high aesthetic standards. But ultimately, when he decided to allow the world to see the finalized work, viewers would stand in awe. Lovers of art can still feel that way today looking at his *David*, *Moses*, the first *Pietà*, and, of course, the Sistine Chapel ceiling fresco. His beautiful and dramatic forms heavily influenced subsequent Mannerist painting and sculpture—besides introducing iconic images into world civilization. Legend has it regarding *The Deposition*, which Michelangelo worked on for eight full years when in his 70s, he destroyed the sculpture in a fit of rage over imperfections in the design or the marble, or both. True or not, the story reflects the views that other artists maintained regarding his perfectionism. No wonder his younger contemporary Giorgio Vasari called him the Divine Michelangelo, like a flawless angel who descended from heaven to create inimitable masterworks.

As Michelangelo's creative career suggests, perfectionism can interfere with productivity—and extremist perfectionism can do so all the more. The French novelist Gustave Flaubert provides a clear example. Famous for ceaselessly striving to find *le mot juste* (the right word), he could spend as much as a week writing a single page, and still remain unhappy with the resulting prose. Hence, Flaubert never came close to the novel-per-year pace of his contemporaries. The American filmmaker Stanley Kubrick's perfectionism permeated nearly all aspects of cinematic creativity, from the exorbitant number of takes that he required of his actors to the total control he demanded over writing, editing, and even music. Although Kubrick would produce, write, and direct some of the best-known films of the latter half of the 20th century—*Dr. Strangelove, 2001: Space Odyssey, A Clockwork Orange, Barry Lyndon, The Shining,* and *Full Metal Jacket*—he averaged only about one film every three years. That's about one-third the output rate of the illustrious Alfred Hitchcock!

Imperfect Genius

Notwithstanding the drawbacks of perfectionism, the potential repercussions of *imperfections* are immense. Flawed works commonly imply flawed genius at best. More truthfully, products that lack genuine perfection might suggest a mere talent whose creativity falls short of genius. So-called one-hit wonders, especially, cannot establish the status of a creative genius unless that single work on which they stake their posthumous reputation closely approximates if not attains perfection. Johann Pachelbel's *Canon in D* may seem a vastly overplayed little piece that somehow ends up in multiple arrangements in both classical and popular music. Yet what probably makes it a stroke of genius is a melodic perfection ideally suited for its constrained form. Could a single note be altered without ruining the sublime impression the canon creates on even the most casual listener?

It's not all that simple, admittedly. For example, what are we to make of the creator whose status as genius is made or broken by a work that can be best described by the oxymoron "flawed masterpiece?" To be generous, such creative products might still earn genius points if there exist extenuating circumstances. For instance, if you google "flawed masterpieces," motion pictures frequently pop up on the screen as examples. This makes sense because even the greatest cinematic auteur most often lacks complete creative control over the ultimate product distributed in the theaters. That reality gives the rationale for the later "director's cut" that supposedly offers the more perfect film, the one that should have been shown had crass philistine studios not gotten in the way.

The Magnificent Ambersons of the American filmmaker Orson Welles defines a case in point. Although Welles unequivocally showed off his genius in the earlier *Citizen Kane*, what would have happened had William Randolph Hearst—the newspaper publisher whose life and loves inspired the film's title character—succeeded in his vigorous attempts to destroy all prints of the scandalous film? Then Welles's claim to genius might have had to depend on the work that came just two years later. Unfortunately, control of that film's editing was taken over by the studio, which then proceeded to mutilate the product—by cutting (and destroying) an hour's worth of footage, for instance, and making the ending happy rather than sad! The film's composer, the famed Bernard Herrmann, actually disowned his contributions in response to the studio's drastic music edits. Despite all of this, if a flawed masterpiece like *The Magnificent Ambersons* was all that remained to defend Welles's reputation, we still might call him a cinematic genius, making allowance for what happened. After all, Praxiteles of Athens is still considered one of the greatest artistic geniuses of antiquity even though none of his sculptures survive except in damaged and often crude Roman copies—such as his well-known *Hermes and the Infant Dionysus*. Just enough has to survive to *infer* creative genius.

Precursor Genius

A totally contrasting scenario emerges when a highly creative individual comes up with a truly brilliant concept that is too far ahead of its time for practical realization. Here we turn to computer science for an illustration. The English polymath Charles Babbage originated the idea of a digital programmable computer, and Ada Lovelace, the daughter of the poet Lord Byron, collaborated with Babbage, publishing a computer program far superior to his (that is, more complex with fewer errors). Yet all of their efforts came to naught. In the absence of electronics, his "computer" had to consist of a complex mesh of gears even more intimidatingly intricate than the clocks and calculators of the time. So the device was never completed. Only much later, when the first electronic computers saw the light of day, could the two collaborators become fully appreciated as the forerunners of artificial intelligence. In 1980 Ada had a programming language named after her, and in the 1990s Babbage's computer was finally constructed to his specifications—and worked! Both can certainly be called *precursor* geniuses notwithstanding the pragmatic imperfections in their programmatic contributions.

Yet wait! Even if I promised back in the prologue to treat readers to ample anecdotes that illustrate the main points, thus far I have not brought in relevant empirical research. It's time to put more balance in the presentation.

Why Not Become Imperfect?

The problem inherent in the foregoing discussion is that absolutely no empirical evidence suggests a strong positive relation between creative achievement and perfectionism as a personality trait. For example, perfectionism is associated with extreme scores on the conscientiousness factor (the "C") in the Big Five personality traits (represented by the acronym OCEAN and mentioned under Tip 3). Yet according to one recent study, conscientiousness fails to correlate positively with various indicators of scientific achievement, quite unlike openness to experience, which again has a positive relation. Indeed, the conscientiousness correlations are actually slightly negative, even if not significantly so. Hence, we cannot even fudge with the lame statement that "the trends are in the right direction." In more concrete terms, scientists who create more high-impact publications are not necessarily more painstaking, precise, deliberate, thorough, organized, planful, reliable, dependable, efficient, responsible, or practical; they are not inclined to be less careless, slipshod, disorderly, irresponsible, undependable, frivolous, or forgetful.

Need a visual representation? Just look at many of the familiar photos of Albert Einstein readily available on the internet. Did you find any showing him with disheveled, uncut hair, sporting baggy clothes and sandals sans socks? One big toe projected well beyond his second, you might also see, thus wearing holes in his socks. *So why bother wearing them*, he then declared. Only if sloppy dress can be done to perfection can Einstein be viewed as perfectly dressed.

Moreover, the empirical case for the perfectionist personality becomes even worse for artistic creators, who score much lower than the norm on conscientiousness. When I tried to dig up a specific instance, the American songwriter Bob Dylan came immediately to mind—particularly his rather lackadaisical attitude when proclaimed the 2016 recipient of the Nobel Prize for Literature. Not only did he delay in acknowledging the honor, but he also procrastinated in writing and delivering his award address, a stipulation for actually receiving nearly a million dollars in hard cash. Even then, he just barely met the deadline by posting his talk on YouTube—plus he has been accused of cobbling together portions of this speech from a SparkNotes.com study guide for Herman Melville's *Moby-Dick*! Whether or not that accusation proves justified, had Dylan been a perfectionist, he probably would have responded very differently to this fantastic honor. Any perfectionism he possessed was undoubtedly applied to the lyrics and music of his songs. Perhaps he, like so many other artists, channel so much perfectionism into their masterworks that nothing remains for more mundane matters, like making conscientious travel arrangements to Stockholm.

So there we have it: scientific geniuses are unlikely to be more perfectionist than the average person on the street, and artistic geniuses are very likely to be less perfectionist than that average—except when it comes to their actual creative output. Fuzzing up the picture even more, we have every reason to argue that imperfection is a necessary component of creative genius.

This imperfection betrays itself not in the final product, but rather in the pathway to a high-impact product, whether scientific or artistic. Ironically, the only permissible path to perfection is imperfection. This fact is evident in two key aspects of creative genius: creativity and productivity.

Creativity: Trials and Errors in Thoughts and Acts

Ada Lovelace's pioneering computer program was an algorithm designed to calculate a sequence of Bernoulli numbers. Although the program was definitely ambitious for the time, the programmed activity itself would not count as a form of mathematical creativity in either her time or ours. Even so, she also speculated on whether computers could be programmed to actually create, such as compose original music. She thus anticipated by over a century Alan Turing's influential 1950 paper on whether "computing machinery" could display intelligence even to the point of successfully passing the "imitation game," or what has later become known as the Turing Test. A computer passes this exam if human judges could not distinguish its conversational skills from those of a live-in-the-flesh *Homo sapiens*. Elevating this question to yet another level, can a computer program be written that creates masterpieces indistinguishable from those produced by world-renowned creative geniuses?

On first glance, the answer might seem affirmative. Wasn't a chess genius, Gary Kasparov, beat by Deep Blue, as I noted under Tip 5? If Kasparov was a genius, why not Deep Blue? One problem with this argument is that creative genius cannot be compared to chess genius. Chess has well-defined rules and goals, whereas genius-level creativity is often distinguished by new and vaguely defined rules and highly novel and even open-ended goals. At the beginning of the 20th century, relativity and quantum theories redefined the very nature of theoretical physics—changes that physicists are still grappling with today. How well would Deep Blue do if the rules and goals were subject to constant flux? Einstein himself never learned how to do creative physics after quantum theory took over his domain, a theory he thought he had logically refuted. Consequently, he devoted the last decades of his life to developing a quantum-free unified field theory that was doomed to fail. The rules had changed.

But what about all of the computer programs that purport to simulate creative genius? For example, one program rediscovers laws in physics and chemistry that made the original discoverers famous, such as Johannes Kepler's third law of planetary motion and Ohm's law of current and resistance. And another program composes new piano pieces that imitate the styles of Johann Sebastian Bach, Ludwig van Beethoven, and Scott Joplin. Yet such scientific rediscoveries and the imitation of already established musical styles wouldn't count as highly creative if done by humans. So why give computers credit for the genius of Kepler or Joplin? Here the mere imitation game proves deficient as a criterion. For a computer to display creative genius, we really must raise the bar: a computer can only attain that elevated status if it achieves eminence like a human genius is compelled to do—such as create perfect works that win Nobel Prizes and then survive the test of time as masterpieces! This touchstone for genius requires much more programming effort before creative computers can become genuinely believable. I'm not a vitalist, so I'm not saying the task is impossible—only that we're nowhere close to getting there.

Why It's So Hard to Program Creative Genius Why can't computer scientists just study creative geniuses very carefully to learn exactly how they think and then write a program accordingly?

Here lies the crux of the issue: there's no such thing as a single process or procedure that can be used to solve every creative problem under the sun. On the contrary, myriad processes and procedures for problem solving have been identified. Here's a partial list of possibilities culled from the research on creativity and genius: divergent thinking, remote association, cognitive disinhibition or defocused attention, mind wandering, intuition, insight, primary (or primordial) process (or "regression in the service of the ego"), dreams and/or daydreams, overinclusive (or allusive) thinking, analogy, conceptual reframing (or frame shifting), broadening perspective, finding the right question, reversal, tinkering, play, juggling induction and deduction, dissecting the problem, "heuristic search" strategies (such as means-end analysis, hill-climbing, working backward, and trial-and-error), plus operations with exotic labels such as Geneplore (generate and explore), and Janusian, homospatial, and sep-con articulation thinking. Even ordinary, everyday thinking can yield great ideas under suitable circumstances.

My apologies to any researcher who feels offended because I may have overlooked their personal favorite process or procedure in this list. But it's really hard to keep track of them all. Possibilities accumulate almost as fast as new researchers enter the fray. And too often investigators will reinvent the wheel, suddenly discovering a "new" process or procedure that considerably overlaps one already included in the inventory but which somehow has an unrecognizable new name! My reader may already have spotted a few examples.

Even deleting any redundant options, a lot remain, so why so many? Because each process or procedure succeeds some of the time—for that was the reason it entered the collection—but absolutely all fail most of the time. Unlike Ada Lovelace's program for generating Bernoulli numbers, a perfect algorithm simply doesn't exist that reliably guarantees a highly creative idea in one go. Therefore, even the greatest creative genius must engage in trial and error, trying out various tactics until one seems the most promising. Worse still, even when the creator settles temporarily on a particular approach, trial and error remains necessary at a more fine-grained level. For example, if the creator attempts to find a good analogy, the next task is to engage in analogical thinking, trying out one analogy after another until a workable one is found. If no effective analogy appears, then it's time to try out yet another tactic. Or perhaps the person might momentarily give up, enter into a reverie, and experience a flash of insight based on some intuitive process. Not all trials have to be conscious and deliberate. Yet trials remain trials. Furthermore, one trial after another can just yield one "error" after another, so there's nothing gained at the end but innumerable failed attempts. I already gave the example of Einstein's long but futile search for a unified field theory. All that labor and nothing to show for it—except ridicule from colleagues. But who said that being a creative genius had to be easy?

Let Picasso Show Just How Hard It Really Is I'm now overdue for an extended illustration of how a preeminently imperfect trial-and-error process leads to creative perfection. I'll pick the 1937 *Guernica*, an undoubted masterpiece painted by the Spanish artist Pablo Picasso. The large mural depicts the horrors of war, as initially inspired by the bombing of a Basque town during the Spanish Civil War. By this time, the artist had already been painting for over 40 years, becoming quite famous in the process. Hence, he certainly had the requisite expertise to perform the task relatively quickly and even effortlessly. He could have started with a basic compositional sketch showing the organization of the key images, and then gradually filled in the details, making fairly minor adjustments along the way. The whole procedure could be compared to a honing process advancing in an orderly progression toward perfection. Very much like sharpening a knife. That honing would be facilitated by the fact that Picasso chose to carry some images over from some earlier work—most notably his 1935 *Minotauromachy*—and adapt them to the new theme. Adaptation is just honing.

But that's not what happened at all. We know that for sure because Picasso kept the sketches he made in route to the final product, assigning each sketch a number and dating enough of them to enable the reconstruction of his thinking. Besides that, a photographer took pictures of the mural during successive states. Upon scrutiny, it becomes patent that the artist relied very profoundly on trial and error, starting with his first compositional sketches and then focusing on individual images within the overall composition. In the former case, whole figures would come and go. For instance, early versions of the mural contained a massive up-thrust arm with clenched fist near the center, an image that completely vanished except for a large lamp where the fist used to be. Even when an image quickly became a permanent part of the painting, it would undergo sundry modifications that were far removed from what we would expect from mere honing. Instead of getting closer and closer to the final image, Picasso would venture into possibilities even more remote, eventually having to return an earlier version that he had previously put aside.

The importance of this trial and error is demonstrated by an empirical study that asked independent judges to rate the similarity of any sketched image to the image that showed up in the final mural. The specific images were the raging bull, the mother

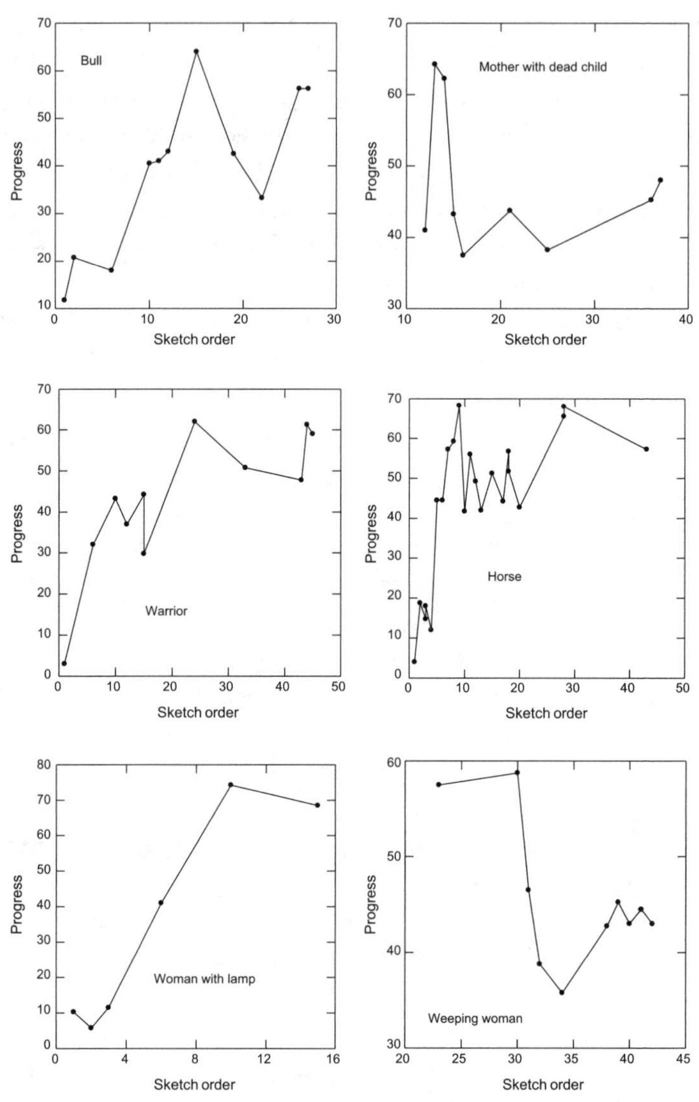

Figure 6.1
Plots of estimated progress scores as a function of true sketch order. Higher progress scores mean that the corresponding sketch is more similar to the final version of the image in the mural.
Source: Simonton 2007b, p. 339.

with her dead child, the dead warrior, the wounded horse, the woman holding her own lamp over the scene, and the weeping woman. Figure 6.1 shows how the similarity ratings vary according to sketch order, where the higher the similarity the greater the progress toward the end result.

It should be obvious that only the woman with the lamp indicates more or less smooth progress toward the goal, except two minuscule dips at the beginning and at the end of the sketch sequence. For the other images, the departures from progressive improvement are so substantial that often a sketch done near the onset of the artist's struggles ends up closest to the accepted version. Indeed, twice (in the sequence for mother with dead child and the sequence for the weeping woman) the second sketch is most similar to the final painting—an outcome that's especially dramatic for the weeping woman. In this latter case, Picasso kept on making faces that were increasingly traumatized, with wildly distorted features and overly abundant tears, until he decided the image was well over the top, obliging him to backtrack to a more restrained depiction of intense grief.

Interestingly, the mural photographs reveal another trial-and-error sequence not seen in the sketches. Some photos show a piece of wallpaper attached to the canvas and some don't. The wallpaper comes and goes, and is placed in different parts of the canvas. What was this all about? Picasso was experimenting with converting the entire monochrome painting into a collage, but ultimately decided against the conversion. So once again, here's an example of how perfection could only be attained by going through an imperfect process. If the process had been perfect, Picasso would have signed off on the finished mural without making so many sketches, or without temporarily pasting wallpaper on the canvas. Trial and error is very inefficient precisely because the so-called errors present a very large proportion of the trials. Yet had not all those errors—Picasso's rejected images—gone through variation and selection, the final outcome would have been mediocre, obliging me to find some other masterpiece to illustrate my point.

Productivity: Hits and Misses in Total Output

The trial-and-error strategy just described has so much importance that it has received many alternative names, such as *generate and test, bold conjecture and refutation, selection by consequences,* and *blind-variation and selective-retention.* What the various expressions all share is one basic tenet: creativity, discovery, or invention demands a big risk—the ability and willingness to fail. Failure is not fun, but it's a prerequisite for success whenever that success means knowing something never known before. Because you cannot venture into the unknown and still know where you are going, you might very well just enter into darkness without finding any light whatsoever. The clear implication of this epistemological necessity is a peculiar paradox: although you cannot achieve success as a creative genius without going through failure, failure cannot guarantee success. As a result, the only way to multiply successes is to multiply failures, at least on the average.

Shall we start with a particular example?

Thomas Edison's Lifelong History of Failures Edison is widely recognized as the greatest inventor who ever lived. By the end of his life, he had accumulated 1,093 patents, a figure not surpassed by anyone until the 21st century. He was also notorious for engaging in overt trial-and-error methods, such as the hundreds upon hundreds attempts to find a commercially viable incandescent light filament. Popular culture may have co-opted the resulting light bulb as the prototypical symbol of a flash of insight, but the light bulb itself originated by much more laborious means. At least the immense drudgery of the generate-and-test strategy succeeded in that case. That wasn't always what happened. About the same time that Edison was working on the light bulb, he also attempted to develop a practical fuel cell that could provide the electric power directly from the source of energy, bypassing the furnaces, steam engines, and dynamos that would normally intervene between a piece of coal and a glowing light. Yet that effort failed miserably—and on one occasion his experiments caused an explosion that blew out the windows of his laboratory!

The net result was a 64-year career filled with spectacular hits and catastrophic misses. Besides the electric light bulb, Edison had invented the phonograph and the movie camera, inventions that quickly spread through popular culture in the form of sound recordings and motion pictures, including the 1903 *The Great Train Robbery* produced by Edison Studios. Yet his more numerous failures were just as impressive, even if far more obscure. In addition to the useless and potentially dangerous fuel cell, Edison patented a flying machine that couldn't fly, built and marketed electric automobiles that couldn't go very far before the batteries went dead, and designed houses built entirely out of cement, and even furnished the homes with furniture, refrigerators, pianos, and phonograph cabinets made out of cement!

Why all the cement? That's another disastrous story. After making many millions from the electric light bulb and its distribution system—the basis for the early Edison General Electric Company that now lacks his surname—the inventor decided to take on a project he thought would make him even more rich and famous: figuring out how to extract iron from the low-grade ores then available in the eastern United States. The venture completely failed, with the minor exception that he mastered the technology behind manufacturing high-quality cement. Bag after heavy bag of the stuff! His financial losses on this project were horrendous—about equal to his entire profits from the electric light bulb. Hence, not every one of his 1,093 patents could be considered hits. And that doesn't even count the more than 500 patent applications Edison filed without success! The US Patent Office didn't just rubber-stamp everything he sent their way.

Other Cases of Highly Successful Failures Naturally, we should not draw big conclusions from a single case. Perhaps Edison is unique. But he isn't. Empirical research on hundreds, even thousands of artists and scientists has established that quality is a positive but probabilistic consequence of quantity. Produce more total output, and more hits tend to be generated, but more misses, too. This quantity/quality relation explains why the English American poet W. H. Auden could say: "The chances are that, in the course of his lifetime, the major poet will write more bad poems than the minor." William Shakespeare may be the greatest poet in the English language, but that does not necessitate that all of his 154 sonnets are of equal quality. As a distinguished Shakespeare scholar once noted, even if some of the sonnets "bear the unmistakable stamp of his genius," other sonnets "are no better than many a contemporary could have written." That unevenness is born out in empirical research, the sonnets varying greatly in how often they are quoted and how frequently they show up in anthologies. Computer content analyses can even detect some of the reasons why the sonnets vary so much in poetic impact. The scholar's judgment is not arbitrary.

The statement that creative genius can produce misses as well as hits should not be taken as saying that the person has suddenly lost all of his or her domain-specific expertise, becoming thereby a mere novice. After all, the same Edison was behind the successful light bulb and the failed fuel cell; the same Shakespeare composed the highly acclaimed Sonnet 30 "When to the sessions of sweet silent thought" (do read the rest of the poem when you get a chance) and the forgettable Sonnet 154 "The little Love-god, lying once asleep" (don't bother reading; it's never quoted and never appears in selective anthologies). Edison had no reason to know that the fuel cell he was striving for was just not feasible at the time. Why would he think so, since world-recognized experts had warned him earlier that a practical electrical distribution system was impossible as well? And Shakespeare was probably just trying to provide a conventional coda to the concluding "Dark Lady" sonnet sequence by adding two similar poems about Cupid. In an analogous way, the final crashing chords of even the best symphonies are seldom profound as the more nuanced movements that led up to them.

Why Failure Is Ensured by What Creativity Means A more basic reason for the hits and misses emerges from the way we define creativity. Although researchers have promoted numerous definitions, here's the one I believe is best: an idea is creative to the extent that it's simultaneously original, useful, and surprising.

Original means low probability; *useful* indicates that the idea is adaptive, functional, meaningful, or appropriate (depending on the creator's goals); and *surprising* signifies that the idea was not a straightforward application of domain-specific expertise—or what the US Patent Office calls the "nonobvious" criterion. This puts creativity in a very different position relative to other forms of expertise. For instance, if a basketball player practices free throws until she can make 90% of the attempts, that's a good thing. But if a scientist publishes the same article multiple times, that's considered very bad form—not just a lack of originality or surprise but also case of self-plagiarism.

Similarly, great artists most often hope to avoid self-imitation because they are constantly endeavoring to maintain their originality rather than appearing to be mere "one shots." Once Beethoven was presented with an enticing offer: A representative of the London Philharmonic showed up at his door with a commission to write a new symphony. Although the pay was very tempting, the Englishman stipulated that the symphony had to be just like one of his earliest symphonies. Yet Beethoven had already completed the Eighth Symphony and was therefore entering the final "Third Period" style that would yield his concluding Ninth Symphony! The composer was so insulted at an explicit request to repeat himself that he shoved the poor emissary down the stairs! Such a symphony would be neither original nor surprising however financially functional.

Sadly, not all geniuses exhibit Beethoven's degree of artistic integrity. When Picasso first received the Spanish Republican government's commission to contribute a painting at the 1937 Paris International Exposition, his first response was to paint yet another rendition of "the artist's studio," an uninspired project that didn't excite him very much but would definitely get the job done. Only the Guernica bombing compelled him to switch themes and thus enthusiastically create a masterpiece both original and surprising. Yet other times Picasso remained content to replicate himself for sufficient cash. Once, when an art dealer asked the artist to confirm whether he had actually painted a canvas bearing Picasso's signature, the response was "it's a fake." But when the dealer protested that he had earlier seen the artist working on that very painting some years earlier, Picasso responded, "I often paint fakes." Such self-replication can reflect the same expertise found in the original work, but "real" expertise also surfaces when highly proficient forgers paint fake Picassos—a crime all too common these days! Imitation may or may not be the highest form of flattery, but self-imitation is decidedly the lowest form of creativity.

But Can You Beat the Odds?

Becoming a creative genius doesn't look that easy, does it? In the first place, you've got to create one or more masterworks. These creative products would presumably meet sufficiently high standards so that at least one might approach, if not attain, perfection—a bona fide exemplary work in the domain. Although you are better off entering the pantheon of geniuses with multiple masterpieces in your portfolio, you can sneak in as a one-hit wonder with a little luck. I already gave the example of Johann Pachelbel who managed to get within sight of the pinnacles of success with his ubiquitous canon. The problem with taking this route to acclaim is that few one-hit wonders were truthfully one-work creators. Pachelbel himself composed more than 500 works, such as diverse compositions for organ and other keyboards, chamber music, and vocal works from arias and motets to masses and *Magnificats*. That seems like a lot of work just to stake his fame on a short piece played at weddings and in the soundtrack of an Oscar-winning motion picture—the 1980 *Ordinary People*.

An alternative route is to write the "Next Great American Novel" or the equivalent, depending on where you live. Then you put all of your creative eggs into one big basket rather than squandering your life in the name of prolific output, almost all of which ends up getting ignored anyway. In fact, some one-hit wonders are novelists known for a single masterpiece of fiction: Harper Lee's 1960 *To Kill a Mockingbird* (particularly if the posthumous *Go Get a Watchman* is discounted as an earlier draft), Ralph Ellison's 1952 *Invisible Man* (albeit he also wrote notable nonfiction essays), and, from England, Emily Brontë's 1847 *Wuthering Heights* (yet she died at age 30, one year after publication, not really getting another chance at fame). Still, this seems a high-risk strategy given the very large number of one-hit novelists who had to publish many misses as well. The 1936 *Gone with the Wind* was not the only novel written by Margaret Mitchell, nor was the 1897 *Dracula* the only novel by Bram Stoker. Same holds for one-hits like Mary Shelley's 1818 *Frankenstein,* Erich Maria Remarque's 1929 *All Quiet on the Western Front,* and Milan Kundera's 1984 *The Unbearable Lightness of Being*. This raises more questions: How can you know when to stop and start resting on your laurels? Can you ever risk doing so? If you're really a genuine genius, would you even want to stop creating?

Some hope might be gained from scientific research on the quantity/quality relation. Yes, the number of hits is a positive function of total attempts, so that misses will increase along with hits. Yet this relation is probabilistic rather than deterministic. That means that exceptions will always exist at either end. On the one side, some creators are mass-producers who generate far more misses than hits, and, on the other side, some creators are perfectionists who produce far more hits than misses. Significantly, the disposition to be one or the other can be relatively stable over the course of the creator's career. Apparently, it's hard to switch from one strategy to the other. The difference between the two types likely depends on both expertise and personality, both of which are highly stable in adulthood.

So if you wish to increase the odds of success with minimal effort, it might seem that the best option is to become a perfectionist! But does that choice really require the least work? If you revise one manuscript umpteen times instead of writing umpteen different manuscripts once, where is the breakeven point? And how do you determine which manuscript should undergo multiple revisions and which should be thrown into the dustbin as an undeveloped sketch unworthy of further improvement? Plus, how many revisions are even required to achieve perfection? At what point do you stop and say, *this* is perfect, time to move on?

Complicating this decision even more is the problem of working in those artistic traditions in which absolute perfection must be avoided at all costs. One example is found in those Islamic artists who are reputed to refrain from perfection to acknowledge that only Allah (God) is perfect. This spiritual restraint is apparently seen in both calligraphy and geometric art where tiny but detectable departures from divine excellence are introduced. In Japanese art, on the opposite end of the world, resides the idea of *wabi-sabi*. Here ceramics, flower arrangements, bonsai and Zen gardens, and teahouse architecture will deliberately incorporate some degree of randomness, asymmetry, or other flaw that reflects the imperfect and transient nature of this world. Even high art is on the threshold of destruction by the law of entropy! Although these aesthetic principles seem so alien to the ideals of the ancient Greeks, they may play a role in the appreciation modern Europeans developed for the ruins of antiquity. The Parthenon and other destroyed buildings on the Athenian Acropolis still draw millions of tourists every year. Even modern artists will create busts and other fragmented depictions of the human form, like a mere disembodied hand.

Given the above problems in attaining either perfection or perfect imperfection, perhaps mass production offers the best strategy. Just operate according to satisficing rather than optimization, so that any attempt at perfection is relinquished. Satisficing just means that each work satisfies minimal standards, without necessarily optimizing all criteria for success. Giving up on perfectionism doesn't mean that you will not produce anything perfect, but rather that perfection will happen from time to time because of the sheer mass of output. Granted sufficient productivity, the likelihood of something perfect or nearly so is bound to happen, even if just by chance. Can Pachelbel's *Canon in D* be cited as an example? Hundreds of attempts and one gigantic hit. Might not *that* be cost effective?

Just Follow Mozart's Example!

Yet why can't you be both a perfectionist and a mass-producer at the same time? Isn't it possible to virtually mass-produce perfection? A creative genius might have such high standards from the outset, and such a superlative expertise, that successes far exceed failures. Almost every thought quickly becomes a polished masterpiece. Thinking of this possibility, it's Wolfgang Amadeus Mozart who seems the potential exemplar. Once we get past his probationary period, as loosely defined by the 10-year rule discussed under Tip 5, the ratio of quality to quantity seems amazing. During Mozart's mature years, somewhere between 60% and 70% of his compositions can be considered hits. Those hits include some of the greatest string quartets, piano concertos, symphonies, and operas in the classical repertoire. Yet during that roughly 20-year interval Mozart created approximately 500 compositions, for an average of about 2 compositions per month! Certainly not much time for polishing gems.

So you *can* beat the odds, just become the Mozart of your chosen domain. Please tell me how that works out for you!

Tip 7
Turn Yourself into a Child Prodigy / Wait Until You Can Become a Late Bloomer!

All pediatricians, and probably all conscientious parents, are familiar with the developmental norms applicable to any human embryo and infant. When during pregnancy does the embryo's heart first start to beat, when do the brain and spinal cord emerge, and when do the other organs initiate their growth? After birth, in what order and at what ages do babies acquire the ability to control their lips, tongues, eye movements, and various external parts of their body, from head to toe? Finally, and perhaps most critical for many moms and dads, when will a baby start to sit up, stand, walk, and even run? When I taught child development in my introductory psychology course, I always showed the class a chart exhibiting the expected order and age in which these benchmarks occurred. Similar charts can be found in any parenting guide as well. But do you even know who first established these developmental norms? A Yale professor named Dr. Arnold Gesell. Unless you're a child psychologist, you may not have heard of him, yet his name shows up very often in biographical dictionaries and encyclopedias devoted to the history of psychology. And deservedly so! Gesell is a prime example of someone who has flown below the radar of popular recognition. And yet, although remembered best among the historians of his discipline, he has certainly left a lasting mark in the landscape of history.

This little snippet of my discipline's history is designed to highlight a key transition point in the *Genius Checklist*. The previous half-dozen tips all concentrated on how prospective geniuses might differ with respect to achievement, IQ, mental illness, genetic endowment, education, birth order, interests, and risk-taking, among other things. But now the preoccupation switches from *who* to *when*. At what chronological age do things begin to happen and at what age do they stop? Does it matter how the end comes? Those are the subjects of both Tip 7 and Tip 8, which explicitly deal with how creative genius emerges and manifests itself across the lifespan. That trajectory can also be tracked in terms of important career landmarks.

So let us begin with the first such landmark—when creative genius really takes off.

Kids in a Big Hurry!

If the certification of creative genius depends on the production of a major work—as separated from apprentice pieces and juvenilia—then becoming a child prodigy seems an ideal way to go. After all, such a person is someone who has attained adult-level proficiency in some domain by the time they are 10 years old, or even younger. By definition, such prodigies can thus circumvent the 10-year rule I discussed under Tip 5. As seen then, when it comes to mastering a chosen creative domain, prodigies not only start earlier than the norm, but also take less time than the norm. Expertise acquisition is thus doubly accelerated. That twofold acceleration should allow the first hit to arrive sooner

Table 7.1
Famous (and Infamous) Child Prodigies Active in Creative Domains

Scientists: Svante Arrhenius, Enrico Fermi, Sigmund Freud, Francis Galton, Carl Friedrich Gauss, William Rowan Hamilton, Ted Kaczynski, John von Neumann, Blaise Pascal, Jean Piaget, William James Sidis, Terence Tao, and Norbert Wiener

Thinkers: Avicenna (Ibn Sīnā), Jeremy Bentham, Hugo Grotius, and J. S. Mill

Writers: Daisy Ashford, Rubén Darío, Barbara Newhall Follett, Samuel Johnson, Thomas Babington Macaulay, Pablo Neruda, Alexander Pushkin, and Arthur Rimbaud

Artists: Basquiat, Gian Lorenzo Bernini, Albrecht Dürer, Mi Fu, Pablo Picasso, and Wang Yani

Composers: Isaac Albéniz, Béla Bartók, Georges Bizet, Frédéric Chopin, Erich Wolfgang Korngold, Felix Mendelssohn, Wolfgang Amadeus Mozart, Sergei Prokofiev, and Camille Saint-Saëns.

during the lifespan—like children who learn to walk or talk noticeably sooner than the established developmental norms.

In table 7.1, I provide some examples of child prodigies.

Some of the names are already familiar. Freud, Mill, Picasso, Mozart, and Saint-Saëns have all been discussed under previous tips. Some of the others have been mentioned in passing as well. Under Tip 1, for instance, I noted that Blaise Pascal has been credited with an IQ of 195, and under Tip 5 I identified him as a phenomenal polymath—with exceptional achievements in technology, mathematics, physics, literature, philosophy, and religion. Now I can add that he was indeed an exemplary mathematical prodigy. While in his late teens, he began pioneering work on calculating machines, work that later led to his speculating about whether such machines have minds, and thus anticipating later speculations in the field of artificial intelligence. But even earlier he'd written an original work on projective geometry. Pascal's contribution was so precocious that René Descartes, the greatest French mathematician at the time, first thought that it was Pascal's father, not the son, who wrote the treatise, still known today as "Pascal's theorem" (also known as the Mystic Hexagram). This first career landmark, which has inspired numerous proofs for the past few centuries, appeared when he was just 16 years old. That's about a decade earlier than normal. To be more specific, the average mathematical genius produces the first hit at age 27. So it seems that status as a child prodigy really does speed up the onset of the career trajectory.

Even more significant is the fact that an accelerated career onset can have long-term consequences. In classical music, for example, the early bloomers tend to produce works at a higher rate, which then results in a higher total lifetime output. Given the equal-odds baseline introduced under Tip 6, that prolific productivity implies a larger number of masterpieces by the end of the career. The cumulative inventory of masterworks then enhances the odds of both contemporary and posthumous acclaim. These assets can even compensate for an abbreviated lifespan. Mozart again provides an archetypical example. Though he died at age 35, he still created well more than 600 compositions, a count more than sufficient to ensure his high place in the classical repertoire.

Ludwig van Beethoven's creative career helps highlight Mozart's achievement. Beethoven was not really a child prodigy (albeit his father tried to sell him as one by lying about his son's age). He was far less productive than Mozart, despite living to age 56—contributing just a little over 100 compositions. Those compositions fall into three periods, his most famous works, such as his immortal Fifth Symphony, coming out of his second period. Yet the works that first proclaimed this phase, such as his Third

"Eroica" Symphony, did not emerge until the composer was approaching 35 years old! So if he had only lived as long as Mozart, we'd know this master almost entirely by compositions that were profoundly dependent on both Mozart and Franz Joseph Haydn. Accordingly, Beethoven would have enjoyed absolutely no claim to be one of the greatest composers who ever lived.

Unhappily, the prodigy route to genius doesn't always work. There are two main reasons. First, this option is not readily available for all domains of creativity. Second, a prodigy's social development may not keep up with his or her intellectual development, with rather crucial repercussions down the line.

Child Prodigies Play Favorites with Domains

Even though the names in table 7.1 do not constitute a representative sample of child prodigies, a quick inspection might support the inference that such human phenomena are more likely to appear in science and classical composition. Actually, in the case of the scientists, almost all—with the conspicuous exceptions of Freud and Piaget—made heavy use of mathematics in their creativity. To understand why, we must also acknowledge that chess geniuses are also highly likely to have been child prodigies. The examples include Magnus Carlsen, Bobby Fischer, Gary Kasparov, Judit Polgár, Samuel Rashevsky, Nigel Short, and Boris Spassky. Carlsen amply illustrates how fast a chess prodigy can go from 0 to 100 mph. He started to learn chess when he was 5 years old, and competed in his first tournament just three years later. At age 12 he demonstrated world-class expertise when he obtained an international master title and one year later earned his grandmaster title. Given that $12-5=7$, not 10, he left the 10-year rule in the dust. Then shortly before age 23, Carlsen became the world chess champion.

If child prodigies seem most visible in mathematics, music composition, and chess, then what do these three achievement domains have in common? The answer should be pretty evident: all three represent highly abstract domains with well-defined and relatively finite rules and goals. This commonality is most obvious in mathematics and chess, but it is even true for the composition of classical music—with its well-delineated rules of harmony, rhythm, counterpoint, and orchestration. It's indicative of their shared conceptual emphasis that all three forms of extraordinary achievement are describable in terms of abstract notational schemes, whether a mathematical proof, a musical score, or chess notation describing a given match. (Nothing comparable exists for domains like literature or the visual arts. Even poetry analysis, for all its abstract depictions of rhyme and metric schemes, falls far short if relied on as an exhaustive description of a given poem.) As a result, mathematicians, composers, and chess players can go from novice to expert in a minimal amount of time. It is no accident that the average age that classical composers create their first hits is around 29, only a couple of years later than eminent mathematicians.

By comparison, creativity in more concrete and complex domains most often requires a longer period of apprenticeship. Indeed, domain-specific expertise must frequently be combined with broad and deep life experiences before a creative career can take off. This requirement explains why it's so very rare for novelists to create their first major work in their teens. The American novelist Barbara Newhall Follett might count as an exception, for she published two novels, *The House Without Windows* and *The Voyage of the Norman D.*, at ages 13 and 14, respectively. Although both works enjoyed some critical acclaim, neither achieved long-term success, nor were they succeeded by later novels even though a full decade lapsed before the author mysteriously vanished from the face of the earth. In contrast, Charles Dickens was a prolific English novelist, but his first major novels didn't start appearing until he was in his mid-20s. Same holds for the Russian novelist Fyodor Dostoevsky and innumerable other ambitious writers. Thus, Follett seems the exception that proves the rule.

An even more strange exception is seen in Daisy Ashford, who is also listed in table 7.1 because of her novella *The Young Visiters*, which she wrote at age 9! Yet this instance of juvenilia is such an epitome of immaturity—replete with spelling errors, grammatical mistakes, and naïve observations about upper-class English society—that it somehow acquired a certain amount of undeserved charm. Who cannot smile at a line like this: "My own idear is that these things are as piffle before the wind." When the book was finally published almost 20 years later, it went through 18 reprintings and eventually became the basis for a play, a musical, a film, and a made-for-TV movie! Yet nobody would argue that this tiny piece could compete with the literary skill and life experience demonstrated in *Oliver Twist* or *Poor Folk*. Hence, it's probably far better not to consider Daisy Ashford a literary prodigy, for her first hit and only hit did not closely approximate adulthood competence, even less literary genius. It is telling that when F. Scott Fitzgerald's first novel, *This Side of Paradise*, came out in his mid-20s, one critic explicitly compared the work with *The Young Visiters*. Ouch!

Child Prodigies Not Precocious in Everything Necessary

Although Ashford probably shouldn't be called a child prodigy proper, she's not the only precocious person listed in table 7.1 who never achieved anything of note in adulthood. Two mathematicians are of special interest.

The first case is William James Sidis. As we saw under Tip 4, James Mill home-schooled his son John Stuart Mill to produce a child prodigy and future genius, and Boris Sidis, a distinguished psychiatrist, decided to do something similar when his son was born in 1898. Father got the training off to a running start by naming his son after the famous psychologist William James, who was his father's friend. The son prospered, becoming proficient in several languages—Boris was a polyglot himself—and mastering higher mathematics while still a child. He entered Harvard College at age 11, thereby becoming the youngest person ever to enroll, and showed his stuff very quickly by delivering a talk on four-dimensional bodies at the Harvard Mathematics Club. Yet once he had left his father's direct supervision, the son's life took a different turn. Gradually losing interest in mathematics, he got involved for a time in politics, and later published works on diverse subjects, becoming something of an ineffectual polymath—the proverbial "Jack of all trades, master of none." He never married, and at age 46 died in relative obscurity shortly after winning a lawsuit against the *New Yorker* for its false portrayal of his later life. In fact, William James Sidis became the subject of legends, many if not most on the malicious side. The popular press just loved to document the failure of a one-time child prodigy, as well as condemn his father's seemingly cruel parenting practices. On the last point the media was apparently right. In time, the son became so estranged from the father that he refused to attend Boris's funeral!

The downward trajectory of Sidis seems more due to failures in social development than in intellectual development. He just wanted to be left alone to do his own thing, and mathematics was too strongly associated with his dad. Another illustration of arrested social development can be seen in the second case, that of Ted Kaczynski, who was born in 1942. Also a promising mathematical prodigy, he experienced some disruptive events that may have helped push him off-course socially. For example, an IQ test administered when he was in the fifth grade revealed an IQ of 167, which led to his skipping grades, an intervention that he himself reported had undermined his peer relationships. Later, having entered Harvard at age 16, he got commandeered into a brutal experiment conducted by the famed psychologist Henry Murray, an experiment that may have left a traumatizing effect on him, such as instilling a deep and enduring distrust of the powers that be. Nonetheless, Kaczynski graduated from Harvard and received a PhD in mathematics from the University of Michigan, where he did award-winning work. Yet after taking his first academic job at the University of California, Berkeley, he rapidly fell apart, neither liking to teach nor willing to interact with students, and soon resigned his appointment. Within a handful of years he led a hermit life in a remote cabin in the Montana woods, where he developed an increasingly radical political philosophy. Less than a decade after resigning from UC Berkeley, he mailed his first bomb, which was followed by more than a dozen additional bombs spread over a 17-year period, killing three and injuring nearly two dozen. Kaczynski first focused his insidious attacks on university people (making him all the more frightening to those of us who were academics back then). In 1978 he placed a bomb that never detonated in the cargo hold of a Boeing 727, which made his crime a federal offense; the FBI called the task force for the investigation UNABOM (short for University & Airline BOMber), thus the media dubbed him Unabomber. Now imprisoned for life, his days as a mathematical prodigy are long over.

I hasten to add that these two cases should not be taken as typical. They just illustrate what happens when social development fails to keep pace with intellectual development. Table 7.1 includes Norbert Wiener, a child prodigy who did grow up to become a famous mathematician—most notable for founding cybernetics. Wiener devotes some pages to Sidis in his autobiography. As a Harvard graduate student, Wiener had been favorably impressed by the presentation Sidis gave at the Mathematics Club. But Wiener also observed how incredibly immature Sidis was, even less mature than the majority of 11-year-olds. Wiener suggests that Sidis could have prospered as a mathematician, had not this immaturity stood in the way.

Just to show how a person can smoothly advance from child prodigy to adult genius, let me end this section with another figure found in table 7.1, namely Terence Tao, born in 1975. As a mathematical prodigy, he was far more precocious than either Sidis or Kaczynski. At age 9 Tao was already taking university-level math courses, and from age 10 to 12 he competed in the International Mathematical Olympiads where he won bronze, silver, and gold medals—the youngest such three-time awardee ever. After earning his PhD from Princeton at age 21, he accepted a position at UCLA, where he was promoted to full professor in a mere three years—the youngest ever to attain that advancement. Publishing more than 300 articles and more than a dozen books, Tao has won abundant honors for his many mathematical contributions, including the prestigious Fields Medal, often considered the "Nobel Prize of Mathematics." He has taught courses at both undergraduate and graduate levels (from linear algebra to Hilbert's fifth problem), collaborated with numerous colleagues, and as a result earned the distinction of earning Erdős number 2 (please google for an explanation of why this number is significant). Tao married a literal "rocket scientist" (an engineer at the Jet Propulsion Laboratory) with whom he parents two children. He thus provides an excellent antithesis to the tragic tales of Sidis and Kaczynski.

So please seek the untroubled path of Tao—if you can!

Taking Your Dear Sweet Time

All child prodigies are developmentally advanced in acquiring sufficient domain-specific expertise to make a big splash at an unusually precocious age. At least that's true if they have initially chosen the right domain, and if their social development doesn't fall too far behind their intellectual development. Yet what about those talents who lag behind the norms rather than lunging ahead of them? What about the late bloomers?

I must start with a clarification. The designation "late bloomers" has more than one meaning. One common usage involves reference to developmental norms applicable to all *Homo sapiens*. Late-talkers provide an example, a group that can encompass anybody, but supposedly also includes the Nobel laureates Albert Einstein and Richard Feynman, along with other geniuses in the mathematical sciences. Yet sometimes the idea extends to a mental state that doesn't directly concern slowed development. A case in point is dyslexia, a reading disability frequently associated with creativity in nonverbal domains, such as the visual arts. Whether justified or not, Picasso is often given as an instance of someone so diagnosed. Yet dyslexia is not a quantitative measure of being early or late relative to some norm, but rather a qualitatively distinct cognitive condition that can be used to explain departures from what's normative, such as difficulties in learning how to read. In a similar vein are diagnoses associated with the autism spectrum disorder, a condition also linked to creative genius in science and math—like Isaac

Newton—as well as chess geniuses and music prodigies. Because autism has a genetic component, it's worth observing that Terence Tao has a younger brother identified as an autistic savant who scored 180 on an IQ test, earned degrees in both music and mathematics, became an international chess master, and got a job at Google Australia. Obviously, Nigel Tao is extremely "high functioning," to use the popular term no longer sanctioned in the psychiatrists' current diagnostic manual.

However, I prefer to avoid using the term "late bloomers" in these ways. It's hard enough trying to accurately diagnose a live-in-the-flesh human being without attempting to perform the same at-a-distance judgment for creative geniuses, particularly those who are already deceased. Any symptoms are far subtler than those I discussed under Tip 2 regarding the "mad genius." For example, Thomas Edison's difficulties in school, which led to his being homeschooled by his mother (as noted under Tip 3), have been ascribed to that old standby, attention-deficit hyperactivity disorder (or ADHD). Yet who really can say for sure? And besides, these factors are often better subsumed under the developmental concept of diversifying experiences introduced under Tip 3. Dyslexia and ADHD certainly qualify as such.

The alternative and preferred usage here is to consider "late bloomers" in a sense restricted to the anticipated developmental progression of a genius's creative career. From that perspective, late bloomers become more or less the mirror image of the early bloomers represented by those creative geniuses who were child prodigies. Instead of starting the acquisition of domain-specific expertise at a much younger age than usual, late bloomers might begin acquisition at an appreciably older age. Whatever the age of this onset, rather than taking considerably less time to achieve mastery, the late bloomers may take considerably longer than normally expected. And as a partial end product of these two contrasts, rather than produce the first hit when relatively young, the late bloomers might create their first hits when rather old.

To comprehend how becoming a creative genius can end up taking longer and thus occur at an older age, we should examine the following topics: crystalizing experiences and life changes, expertise acquisition on the slow track, and the vicissitudes of one-hit wonders.

Crystallizing Experiences and Life Changes

Child prodigies are extremely lucky. Somehow they manage to discover an intense interest in mastering a demanding domain at a very young age. If you asked preschoolers "What do you want to be when you grow up?" most would not say, "I want to become a famous mathematician!" At best, they might mention firefighter, rock star, or President of the United States. Moreover, if you wanted the kid to elaborate on the choice by specifying what needs to be done to get to that desired destination a couple of decades down the road, don't expect a detailed answer, if any. For most kids that age it's enough just to eat and sleep. Yet child prodigies could tell you how many hours a week they spend taking lessons and practicing, plus demonstrate how well they have progressed in their studies.

To be sure, even child prodigies can change their minds, at some time deciding to switch domains. Yet when they do so, the new and old domains are seldom drastically different. A would-be piano virtuoso in the classical repertoire might slide into classical composition or start work on jazz improvisation. Although classical and jazz expertise are not equivalent, whether in performance or composition, they overlap enough for valuable positive transfer to take place. For example, the American jazz great Herbie Hancock began as a piano prodigy in classical music. At age 11 he even performed a movement from a Mozart concerto with the Chicago Symphony Orchestra at a young people's concert. Hancock didn't begin formal training in jazz until he was 20 years old. But it's not like he had to start all over again from scratch. On the contrary, within a mere two years he came out with his first hit album, appropriately titled *Takin' Off*. All of the compositions were his own, and the first track featured his "Watermelon Man," which soon became one of the all-time jazz standards. Learning to play the first movement of Mozart's Coronation Concerto with a major classical orchestra didn't seem to impede Hancock's emergence as a jazz artist.

Creative Epiphanies A teenage encounter with recordings by The Hi-Lo's initiated Hancock's shift in focus from classical to jazz. Yet hearing a live performance by jazz pianist Chris Anderson was what finally convinced him that he needed formal training, so Hancock asked Anderson to take him on as a student. This event illustrates the so-called crystallizing experience. Before the performance, Hancock was a talent searching for its genius. Considerable amount of musical expertise had accumulated, but it had to be channeled into the right domain, a domain close enough to his classical training that he would not need to throw everything away. Suddenly all came together and, as they say, the rest is history.

Crystallizing experiences happen fairly often. One empirical inquiry identified possible examples in the classical composers Richard Wagner, Claude Debussy, Igor Stravinsky, and Pierre Boulez; in the mathematicians Carl Friedrich Gauss, Évariste Galois, David Hilbert, G. H. Hardy, and Srinivasa Ramanujan;

and in the visual artists Pierre-Auguste Renoir and Paul Klee. In each case, some distinct event either sets talent development on a totally different course or launches talent development in the first place. Happily, in these cases the crystallizing experience occurred soon enough that the creative geniuses could avoid becoming late bloomers. All could claim either normal or precocious development. Yet sometimes the precipitating event happens much later, thereby yielding a much-delayed creative trajectory.

The Austrian composer Anton Bruckner offers a prime example. Not a child prodigy by any means, he received some ordinary music training in school, learned to play the organ, and obtained experience as a choirboy in a monastery. By his late teens had become an excellent organist, and had already begun his first compositions, almost exclusively for the church, and nothing truly outstanding. So he continued his hardworking education with various teachers, still not thinking about venturing far beyond religious music as his main vehicle of expression. Then bursts in the crystallizing experience: one of his instructors introduced him to Wagner's new opera music. Although Bruckner was 39 years old at the time, and could be excused if he had he already been set in his ways, he still switched his emphasis to symphonic music. Because Wagner's compositions were much more progressive than the conservative church music that Bruckner had been composing for more than two decades, it was almost like changing careers—like resetting the clock on the 10-year rule! Consequently, his first symphonies were not successful, especially the Study Symphony and Symphony no. 0. Not until his Symphony no. 3 did he produce one deemed good enough to allow Wagner to accept the dedication (hence the Wagner Symphony). But Bruckner was still not there yet. Finally, he completed the first version of his Symphony no. 4, known as the "Romantic," which became his first big hit, a work still frequently performed and recorded today. By then the composer was 50 years old, meaning that this key career landmark appeared more than 20 years later than the average for classical composers in general—beyond doubt certifying Bruckner as a bona fide late bloomer.

Making Lemonade from Life's Lemons The Bruckner and Hancock cases entailed unexpected musical experiences that altered their musical development. Yet sometimes an adverse change in everyday life will provide the impetus for an unanticipated shift in creativity. Here the striking instance is Anna Mary Robertson "Grandma" Moses, probably America's best-known folk artist. Admittedly, not everyone will agree that folk art represents a genre that attracts creative genius, but her story remains instructive nonetheless.

For most of her adult life, the main artistic interest was embroidering quaint pictures with yarn. Occasionally she would delve into another artistic medium—such as the landscape she did on a fireboard when she was in her late 50s (using regular house paint no less)—but embroidery needles were her tools of choice. Then, in her 70s, she developed very severe arthritis, and by age 76 she found embroidery too painful to pursue. A sister recommended that she return to painting, and thus a remarkable new creative career began. Her art, quite original in style, soon caught on. Though her first paintings only sold for a few bucks, they would eventually go for several thousand. Her *The Fourth of July* hangs in the White House and was featured on a 1969 US postage stamp. Although it's difficult to pick out the first genuine hit from the 50 or so paintings this prolific painter produced each year, Grandma Moses's *Sugaring Off* might count—about a decade ago the painting brought more than a million at auction! That work was created in 1943 when she was 83 years old—making her the late bloomer par excellence!

Expertise Acquisition on the Slow Track

Now we have to go back to the 10-year rule and think about what it really means: To attain world-class expertise in any creative domain supposedly requires an average commitment of 3 hours per day for 10 years, or very roughly 10,000 hours total. Naturally, that time commitment is easier made than done, even when having talent might accelerate the process. This's why becoming a chess grandmaster, violin virtuoso, or star athlete is so hard. Fortunately, many domains of creativity subsidize expertise acquisition by making the domain an academic subject. Thus, future creators in mathematics and the natural sciences can start their training in high school and continue it in undergraduate and graduate education, so that the requisite 10 years follow naturally from lectures, problem sets, exams, labs, and mentors. Anyone with a PhD from top-flight schools, colleges, and universities should be all set to take off. Indeed, the research done for the doctoral dissertation may become the first hit. Should work out well for most talents—unless you face the impediments often imposed by gender or class.

Obstacles of Gender Marie Curie's thesis work on radioactive substances helped her win a share of the Nobel Prize for Physics in 1903; surely, then, we could say she was on the fast track. But in truth, Curie (née Skłodowska) must be viewed as a late bloomer. After all, she was 37 when she completed her initial research on radioactivity, which puts her roughly a half-dozen years behind the average age that physicists produce their first high-impact contribution. Her path of discovery, however, illustrates ways in which the acquisition of domain-specific expertise can be decelerated by extraneous circumstances in one's personal life. In her case, the deceleration was caused by repeated interruptions in her formal training. These occurred in part due to family financial problems as well as gender discrimination in her home country of Poland. It was not until she was near her mid-20s that she could move to France and enroll at the University of Paris. Even then, she was very poor, having to make ends meet by tutoring in the evenings, and often suffering from hunger and the winter cold. So it wasn't until she was 27 that she finished her undergraduate education.

She didn't at once advance to graduate school either, but instead obtained some research experience and visited her family in Warsaw. During this interval she met Pierre Curie, who encouraged her to seek a doctoral degree just as she encouraged him to write up his doctoral work to receive his own PhD, which he did the same year they were married. Their marriage was a very happy one, and they even collaborated on the research leading to the Nobel (which they shared, along with Henri Becquerel).

The couple decided to start a family before Marie obtained her doctorate. She was in her late 20s when her first daughter was born (the Irène discussed under Tip 3), and to support the addition Marie was obliged to take on a teaching job. In addition, the Curies received no research funding from the university, and were even forced to run their lab in old decrepit shack. Worse still, her dissertation topic was extremely time intensive, requiring that tiny quantities of radioactive elements be extracted from tons of raw ores. By now it should be obvious why Marie's first hit came so late!

Marie Curie's status as a late bloomer was partly assured by her gender. The progress of eminent women through formal education way back then was often less rapid than that of eminent men—if the women were even allowed to get advanced training in the first place. For example, a study of illustrious American psychologists born between 1842 and 1912 found that women received their doctorates a half-dozen years later than the men. And that statistic assumes that Mary Calkins—almost Marie Curie's exact contemporary—can be credited with earning her PhD at age 32. By that time Calkins had satisfied the spirit even if not the letter of the academic requirements at Harvard University. But she was denied recognition for her stellar achievements in what today we would call a catch-22. Although Harvard made an exception and allowed her to audit lectures on its campus rather than study at the Annex (the precursor to Radcliffe), it made no exception to the rule that refused to award degrees to women because they were not official Harvard students—even though Calkin's mentors signed off on her doctoral thesis with highest honors, and one of those highly supportive mentors was no other than the psychologist William James!

Socioeconomic class can also exert an adverse impact on expertise acquisition, yielding another type of late bloomer.

Obstructions of Class Here's an especially intriguing story about a person couldn't start his formal training until *after* he'd already contributed his first hit! That do-it-backward genius was George Green, an English physicist and mathematician whose potent ideas still permeate modern mathematics and physics, including quantum field theory. Born to a baker and a miller, he was sent to school for only a single year before his father withdrew him to engage in fulltime drudgery. At most, that brief attendance would have exposed the boy to a little algebra, trigonometry, and logarithms. Yet he apparently acquired a deep love for mathematics, engaging in self-study without any guidance from teachers, but taking full advantage of books available at a local subscription library. Amazingly, instead of learning the old-fashioned Newton-encumbered mathematics used in Great Britain those days, Green mastered the leading-edge Leibniz-inspired mathematics favored on the European continent. Meanwhile, when he turned 30, growing family responsibilities (for an eventual seven children!) augmented the heavy labors at his father's mill.

Nevertheless, in 1828, at the age of 35, he published a paper at his own expense that introduced a highly original mathematical analysis of electricity and magnetism. Even though it took some time for his contribution to become widely known—Einstein more than a century later remarked that Green's work was 20 years ahead of its time—this late-blooming creative genius gradually received due recognition for his very first hit. The timing was inadvertently convenient too, for Green's father died one year later, leaving an inheritance large enough for him to quit working at the mill and consider extending his mathematical studies at the university level. At age 39 he entered Cambridge, and he graduated with honors six years later. Because graduation happened a decade after his first career landmark, Green might be said to have lived the 10-year rule in reverse—if it were not for all of his hard work studying on his own. It is fitting that the mill that Green once so despised has since been converted into a science center.

Vicissitudes of One-Hit Wonders

In the previous two sections I have tried to explain the occurrence of the late bloomers. In the first, late bloomers have to dilly-dally for a while before some key event, a crystallizing experience or life change, causes them to grasp their real calling. This career insight may even require some degree of retooling because the expertise they have already acquired may not be quite up to the task. In the case of Bruckner, writing for organ or choir does not carry over well to composing lush orchestral symphonies eventually containing big Wagner tubas instead of a keyboard or voices. In the second section, late bloomers just take longer than average to completely acquire the required domain-specific expertise. External conditions merely got in the way. Both Marie Curie and Green knew their passions, and stuck to them throughout their lives, but there's only so much you can accomplish when work or family gets in the way.

No doubt both of these explanations have some validity, depending on the creative genius to which they apply. Indeed, both can operate at the same time for a given creator. This joint action is probably commonplace in those who change creative domains. Unless the person has attained financial independence first, expertise acquisition on the second domain will have to take place while continuing work on the first domain. So a lengthened preparation period is added to the late start. That said, the two phases of the career might occur in sync so that creativity continues unabated.

Take William James, who came up a lot under Tip 4. He started out in physiology, then jumped to psychology, and finally capped off his creative career in philosophy—with primary emphasis on the last two domains. Admittedly, the transition across these domains was facilitated by the context. Psychology as a discipline had relatively recently emerged out of physiology, and psychology back then was often found in the same academic department as philosophy. Still, the areas of expertise remained sufficiently distinct as to warrant separate labels. In any event, careful examination of James's publication list shows that no precise temporal divide exists between these phases of his career. For example, his 1896 *The Will to Believe*, which is manifestly philosophical in nature, was published well before his 1902 *Varieties of Religious Experience*, which has a clear psychological emphasis. Thus, his psychological and philosophical contributions overlap sufficiently to avoid any lull appearing in his creative career. James never had to take time off to retool.

In any case, both the delayed start of expertise acquisition and the slowed progress of that acquisition can explain the emergence of late bloomers.

Two Wildly Different Routes to One Hit Having just said that both explanations can be valid, I now want to argue that neither may be valid, at least in certain circumstances. This argument depends on what we learned under Tip 6 about quality, or "hits," as a probabilistic function of quantity, or "attempts." The more attempts, the more hits, on the average, but also the more misses, again on the average. Yet we also observed that some creators might consistently exhibit higher hit rates, while others show lower hit rates. Also under Tip 6, I introduced the concept of the one-hit wonder—someone who has but a single big success in an entire lifetime. Putting these two principles together, we see that there are actually two divergent ways of being a one-hit wonder. At one extreme, the creator makes a single attempt, but is unbelievably lucky, for that one attempt yields a hit rather than a miss. For example, the American novelist Norman Maclean wrote just one novel, his 1976 *A River Runs Through It*, which was made into a successful movie. At the other extreme, the creator makes multiple attempts, but is incredibly unlucky, and all but one attempt yields a miss rather than a hit. For instance, another American writer, Upton Sinclair, wrote many novels—almost a dozen in his *Lanny Bud* series alone—but he created only one notable hit, his 1906 *The Jungle*. Naturally, most often the actual case will lie somewhere between these two extremes, but the ultimate outcome is the same, one measly hit. Because that single success necessarily defines the first major contribution, is it mandatory that this first hit appear 10 years after the onset of expertise acquisition, at least as an approximation?

The answer is negative. The single success of a one-hit wonder could occur randomly anywhere in the career. This expectation holds regardless of the number of attempts. The necessary consequence is that a late-blooming one-hit wonder is simply a creative genius who might have had to wait longer than average before getting the first hit—*just by chance alone*! If you roll a die with the goal of getting a six, and six symbolizes a hit, then there's only a 17% chance of getting a six on the first roll, and it's very conceivable that it will take multiple rolls to get the hit. In this view, the earlier bloomers among the one-hit wonders do not differ from the later bloomers, except in having greater luck in getting a hit right near the close of the 10-year probationary period.

The Case of an Unlucky Late Bloomer? In Tip 8 I revisit the probabilistic logic I just offered, but right now a reader might ask: Can you give a specific example, like you promised? No, that's not really possible. The effects of chance can only be determined when you have enough cases to work with. For instance, you can't determine if a coin is biased by a single flip. It's got to be flipped multiple times: the smaller the bias, the more trials necessary for bias to be detected. Hence, the best I can do is to provide a purely hypothetical example based on a real person. That person is the Greek-French mathematician Roger Apéry, who died in 1994.

Who? That was my first response, too. I just obtained his name by doing an online search for a one-hit wonder whose single hit appeared toward the very end of the career. Roger Apéry best qualified. When he was 63 years old, Apéry published the first proof that a certain well-defined number was irrational (i.e., could not be expressed as the ratio of two integers, like π can't). For our purposes here we don't have to know *why* this finding was important, but it *was* and remains so (besides being quite interesting, if you're very fond of the Riemann zeta function). When he first gave a talk presenting his proof before his peers, he wasn't believed. But his result was later confirmed, and new proofs added, so that his finding has become known as Apéry theorem and the actual value of the irrational number Apéry's constant (defined, of course, as an infinite series). Sadly, by this time Apéry had already been diagnosed with Parkinson's disease, so the one hit was to prove to be his last. Yet his discovery remained so valuable to him that he had it inscribed on his tombstone—the theorem, that is, not the proof!

So now, for the purposes of my argument, let us suppose that Apéry's proof did not depend on anything he had published before. I realize that's a big assumption, but we're talking about a hypothetical anyway, putting this mathematician in an alternative universe for convenience. In partial defense, I did read the biography of Apéry written by his son, who is also a mathematician, and found no mention that this eponymic theorem represented the culmination of his father's previous results or even that it ensued from another of his findings. Given my assumption, his first and only hit could then have appeared anywhere in his career, including at the beginning, perhaps converting this mathematician into an early bloomer. Or maybe a middle bloomer! See the point? Without knowing the nitty-gritty causal details, we cannot rule out the possibility that late bloomers were just the victims of chance. Apéry merely rolled the die until he got a six, and it turned out that a six only popped up when he was in his early 60s!

Is Normal Optimal?

So what should *you* do? Rush to become a child prodigy? Or stand aside until you know you've finally discovered the true you—your genuine interests and talents—and proceed from there? Pros and cons exist on either side, for sure. If you want to follow the prodigy route, you have to pick your domain very carefully, and do so well before your 10th birthday (so it's already too late for most if not all my readers). You also have to worry about whether your emerging social skills can keep up with the progress in your precocious expertise. If you decide on the late-bloomer path, you'll definitely become mature enough by the time your career takes off, but the patience needed until you get to be a creative genius may be too much to bear. Plus, you may have a day job that keeps you from realizing your dreams. Therefore, adopting the mean between the extremes may be the best choice. After all, the middle path is that chosen by the majority of creative geniuses. They represent the norm, like the average age to walk or talk. The expected age at which creative geniuses make the first high-impact contribution to their chosen domain is somewhere between 25 and 30 years old. So that seems a reasonable developmental benchmark. If that worked for Einstein, why not for you?

But is there anything else we haven't yet examined that might tip the scale toward either side of this Golden Mean? Yes, and the missing consideration is a big one: How long to you plan to live?

Tip 8
Do Your Best to Die Tragically Young / Just Live to a Ripe Old Age!

However the creative career may be analyzed, death is the developmental landmark that's the most certain—just as certain as taxes, according to Benjamin Franklin's famous quip! Whatever goes on after a creator's first big success, another success cannot possibly be created after life ends. To be sure, a posthumously published work might earn the creator widespread acclaim, but that later recognition is most often a function of creativity that happened during the creator's life. For instance, *The Way of All Flesh* was published in 1903, one year after the English author Samuel Butler's death, but he had written this classic novel much earlier; he just dared not publish the book in his lifetime because of its severe criticisms of Victorian mores. In a similar vein, E. M. Forster's *Maurice,* a novel he first drafted in 1914 and revised as late as 1960, was not published until a year after his death in 1970—because Forster thought that the world would not yet accept a book about same-sex love. Admittedly, on rare occasions an editor or literary executor can add value to a work by revising (or excising text from) an unfinished manuscript. One notorious example reveals that the American writer Thomas Wolfe's 1940 novel *You Can't Go Home Again,* published two years after his death courtesy of Edward Aswell, raised a perhaps unresolvable debate about whether Aswell should be considered a creator rather than an editor. To the extent that this novel's success is due more to Aswell than to Wolfe, the latter's score as a creative genius must be docked a few points.

So, considering the "givens"—that death is the inevitable concluding landmark of the creator's actual life, and that the creator's first hit (as I discussed under the Tip 7) is the first career landmark—now it's time to consider what other landmarks we should discuss between the first major contribution and the termination of the genius's whole creative enterprise.

First, Best, and Last Hits

In the case of the one-hit wonders, the story ends as soon as it starts. No successes follow that big one, so the first hit is also the last (and the best) hit. The creator's entire reputation is staked on a single major contribution to a creative domain. That said, creative geniuses of the highest order seldom rest their long-term fame on a single masterwork. Those who are credited with at least three such contributions, in particular, can be characterized by three core career landmarks: the first hit, the best single hit, and the last hit. The two-hit wonders will necessarily merge either the first two or the last two landmarks. The best will be either the first or the last of the pair, with the first more likely to be the best than the last. A well-known illustration in the realm of opera is seen in the career of the Italian composer Pietro Mascagni. Although he created well over a dozen operas, he had only two enduring hits, his 1890 *Cavallieria rusticana* and the 1891 *L'amico Fritz*. The former is far more frequently performed and recorded—and even dominates the soundtrack of the 1990 film *The Godfather Part III* in which the opera becomes a big part of the plot! Hence, *Cavallieria* is both the first hit and best, whereas *L'amico Fritz* brings up the rear as the last hit, albeit only one year later.

To understand better how this all works, let's return to an important principle I first presented under Tip 6 and then repeated under Tip 7: quality is a probabilistic consequence of quantity. The more attempts, the more hits, but also the more misses, on the average. Although this idea was first applied to the hits and misses across various careers, it can also be applied to the fluctuations of his and misses *within* an individual career. Those periods of the career in which a creative genius makes the most attempts will tend to be those in which the most hits (and misses) appear. In contrast, when quantity languishes, so will actual quality.

Figure 8.1 depicts the typical outcome.

The curve starts at age 20, the hypothesized onset of the career. Productive output increases rapidly up to a peak at roughly age 40, and thereafter a gradual decline begins. At the beginning the creator will accumulate many attempts until the first hit finally appears somewhere between age 25 and 30. Similarly, at the end, the appearance of the last hit is contingent on accumulating enough attempts until the chances finally work in the creator's favor between age 50 and 60. Between the first and last hit is the best hit, which has the highest likelihood of occurring where output maximizes (or, more accurately, a little after that peak because of the asymmetry of the distribution). Naturally, once the creator dies, the whole process comes to an abrupt halt. The labors of genius on this earth are done.

To give real stats, let's return to the empirical study of the 120 most eminent composers in the classical repertoire, which we

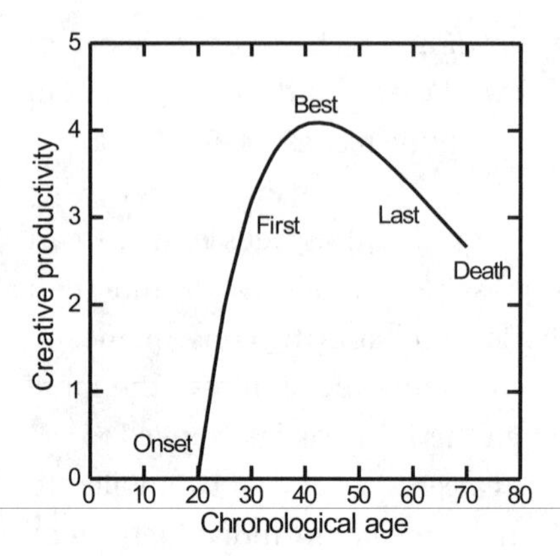

Figure 8.1
The typical relation between chronological age and creative productivity. The overall curve is characterized by career onset, the first hit, the best single hit, the last hit—and death, thus terminating creativity.

first encountered under Tip 5. The average ages corresponding to first, best, and last hits are 29, 40, and 51, respectively. These hits all represented enduring contributions to the standard repertoire. Significantly, the mean age for maximum annual output was 39 years old, indicating the high point of the curve in figure 8.1. Finally, the mean life expectancy was about 66 years. Even for this highly distinguished group, the last 15 years of life do not necessarily yield an enduring contribution to the repertoire. That's a point worth remembering!

The curve describing the change in output with age is based on a very large number of empirical studies involving hundreds, even thousands of highly creative people. For that very reason, the curve does not dictate the career of any one creative person. It's a statistical average only. Hence, the expectation derived from the curve should not be considered deterministic. Consider the curve only as an expression of a probability. Of course, the same holds for the expected locations of the first, best, and last hits on the curve. Likelihoods only, with plenty of room for exceptions.

Indeed, the Mascagni example above provides an obvious outlier. His first and best opera came out when he was 27, his second-best and last hit when he was 28. Despite the fact that Mascagni continued composing, producing his last opera at age 72, his creative career was long over. Some of his fans will point to *Iris* as a worthy opera, but that doesn't help him much given that he was 35 years old when he composed it—still rather early for a last work given someone who lived into his early 80s. No wonder that he reportedly complained later in life, "It is a pity I wrote *Cavalleria* first, for I was crowned before I became king."

Nonetheless, the science of genius is not immediately interested in such idiosyncratic exceptions but looks instead at the systematic departures from the graphic pattern shown in figure 8.1. Given a sufficiently large enough inventory of these orderly departures, we may be able to explain the vast majority of career trajectories that creative genius displays. Happily, investigators have been studying the relation between age and creative productivity ever since 1835, making it the oldest research topic in the science of genius. That first inquiry was conducted by no less a scientist that the Belgian polymath Adolphe Quetelet, who is much better known for establishing not only the normal curve as the basis for understanding how people vary but also the closely connected BMI, or "body-mass index" (and thus the bane of those who find themselves on the upper end of *that* bell-shaped curve). Since then many other researchers have investigated the age-creativity relation, including me, who published my first paper on the question in 1975. What this all means is that we probably know more about this subject than any other discussed in the *Genius Checklist*. That knowledge provides a wealth of potential explanations for lawful departures from the expectations shown in figure 8.1. These explanations involve the specific domain of achievement, individual differences in lifetime productivity, the age at career onset, and differential hit rates.

From Youth-Welcoming to Maturity-Favoring Domains

Although research on the relation between age and creative productivity began with Quetelet, the psychologist who really put the subject on the map was Harvey C. Lehman, whose 1953 book *Age and Achievement* summarized what was known by the middle of the 20th century. One repeated finding was that the optimal age for producing a major creative hit depended on the domain of achievement.

Career Trajectories in Poetry and Prose For example, Lehman found that "most types of poetry show maxima 10 to 15 years earlier than most prose writings other than short stories." Thus, where esteemed novels, including best sellers and critically acclaimed "best books," normally appear between 40 and 44 years of age, memorable odes, elegies, sonnets, lyrics, and other poetic forms more commonly emerge between 25 and 29 years. Nor is this contrast unique to European literatures. My 1975 investigation showed that the age difference was a cross-cultural and transhistorical universal!

Indeed, the age gap between the creative peaks of poet and novelist is wide enough that poets can actually exhibit an appreciably lower life expectancy. Whereas poets can die before they turn 40, and even before they turn 30, and still produce poetic masterpieces, it is much more difficult for novelists to die young and still leave a fiction masterwork behind. This effect just represents a specific case of a more general phenomenon: increased precocity is associated with decreased life span. The English Romantic poets provide ample examples of early mortality: John Keats died before his 26th birthday, Percy Shelley died at age 29, and Robert Burns passed away at only 37. Perhaps the novelist most comparable to these instances is the Russian Mikhail Lermontov, "the poet of the Caucasus," whose psychological novel *A Hero of Our Time* was published the year before his death in a duel at age 26. That admitted, Lermontov's work is relatively short, especially when compared with the fiction masterpieces of Dostoevsky and Tolstoy. In fact, because the novel essentially consists of five short stories, it may still fit Lehman's overall generalization.

Career Trajectories in the Sciences Lehman also observed that the career trajectories displayed major differences among various scientific disciplines. These contrasts have been replicated and extended in a study that I published in 1991, based on more than 2,000 eminent scientists. The main results are presented visually in figure 8.2.

In some respects, the results echo those found for the 120 classical composers. The first hit appears somewhere around the 30th year and the best hit around the 40th year. But the last hit hovers around the mid-50s, and the life expectancy around 70 years, indicating that scientists enjoy somewhat better creative and biological longevity. Another important point is that the four jagged lines do not pursue exactly the same horizontal path. For instance, although the first hit of an astronomer tends to come at an older age than the first hit of a biologist, the age

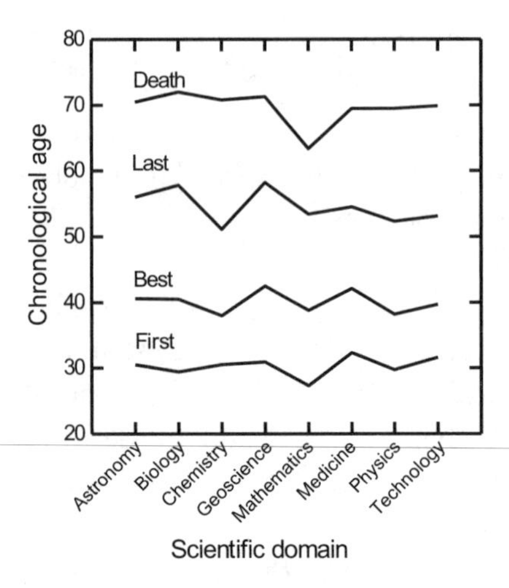

Figure 8.2
The expected ages for first, best, and last hits plus death for eight scientific domains.
Source: Constructed from data reported in Simonton 1991a, table 2.

at best hit is pretty much the same for the two domains, and the last hit appears at an older age for the biologists than for the astronomers.

Yet the eminent mathematicians exhibit considerable consistency: their first hit comes at the youngest age of all domains, their best hit and last hit arrive almost as young as it does for the chemists and physicists, and, most remarkably, their life expectancy is much lower than seen in any other domain—about a half-dozen years earlier than everybody else. That number falls almost exactly in the middle of the the mean life spans of poets and novelists. Hence, mathematicians can be compared to poets. In particular, pure mathematicians (as distinguished from applied mathematicians) appear to show career trajectories that closely parallel those of lyric poets (as distinguished from epic or dramatic poets). To offer a list of mathematicians whose brief lifespans certainly rival those of the Romantic poets cited earlier, Srinivasa Ramanujan only lived to 32, Niels Henrik Abel to 26, and, Évariste Galois to 20, his life cut tragically short, like Lermontov's, in his own ill-advised duel. The very idea that someone as young as 20 could have solved a major problem in algebra that had stymied mathematicians for more than three centuries is truly impressive!

Conceptual versus Experimental Domains The contrast between pure and applied mathematics suggests that a finer distinction can be introduced within domains. To illustrate how, the American economist David Galenson contrasted two contrary creative life cycles. On the one hand, conceptual creators peak early in the course of their careers because they conceive their best works via sudden intellectual breakthroughs. They can also be called "finders" or "young geniuses." Galenson identifies the Spanish artist Pablo Picasso as a prime representative of this type. On the other hand, experimental creators work more gradually via exploration, and thus their best works do not appear until relatively late in life, after they have accumulated a massive amount of expertise. They can also be called "seekers" or "masters," with the French artist Paul Cézanne exemplifying the experimental creator, according to Galenson. Galenson initially presented this life cycle contrast with respect to painting, but the economist eventually extended the idea to include the principle forms of artistic creativity, such as poetry, novels, sculpture, and film. In fact, the scheme was even broadened to encompass scientific genius.

Nonetheless, like so often happens when theorists get too enamored of their theories, Galenson went a step too far when he argued that the distinction between poets and novelists was nothing more than another manifestation of the contrast between *conceptual* and *experimental* creators. More specifically, poets are far more likely to be conceptual creators, and novelists experimental creators. Therefore, the age contrasts between the two literary genres may actually reflect underlying difference in creative life cycles. Once the latter is controlled, the former should vanish.

It's useful first to clarify the criteria Galenson applies to those two categories. For instance, as he sees things, conceptual poets typically emphasize ideas over emotions and imaginary figures and settings over visual images and observations, whereas experimental poets write based on real experiences and perceptions of the external world. In a similar vein, conceptual novelists begin with ideas and abstract facts, whereas experimental novelists emphasize ever-developing real-life experiences. Interestingly, where the former see most of their creativity occurring before they actually start writing, the latter view their creativity as a possess that unfolds during writing.

The main problem concerning Galenson's argument is that his own data fail to support it! A reanalysis of these data showed that poets can exhibit either life cycle, and novelists can do the same. Moreover, according to his data, the two life cycles are almost equally distributed across the two genres (i.e., independent or "orthogonal"). That means that four different creative life cycles exist: conceptual poets, experimental poets, conceptual novelists, and experimental novelists. Each of the four feature a different expected peak in which they are most likely to generate their single biggest hit. The results are shown in table 8.1. An additional statistical analysis shows that even though the experimentalist poets and novelists produce their best work about 11

Table 8.1
Fourfold Typology of Creative Life Cycles in Literature: Expected and Actual Career Peaks

	Poets	Novelists
Conceptualists (finders)	*Expected: 28*	*Expected: 34*
	Eliot (1888–1965): 23	Fitzgerald (1896–1940): 29
	Cummings (1894–1962): 26	Hemingway (1899–1961): 30
	Plath (1932–1963): 30	Melville (1819–1891): 32
	Pound (1885–1972): 30	Lawrence (1885–1930): 35
	Wilbur (1921–2017): 34	Joyce (1882–1941): 40
	Williams (1883–1963): 40	
Experimentalists (seekers)	*Expected: 38*	*Expected: 44*
	Bishop (1911–1979): 29	James (1843–1916): 38
	Moore (1887–1972): 32	Faulkner (1897–1962): 39
	Lowell (1917–1977): 41	Dickens (1812–1870): 41
	Stevens (1879–1955): 42	Woolf (1882–1941): 45
	Frost (1874–1963): 48	Conrad (1857–1924): 47
		Twain (1835–1910): 50
		Hardy (1840–1928): 51

Source: Adapted from Simonton 2007a, table 2. Please note that the above placements are Galenson's, not the author's. Many experts would challenge one or more classifications. Indeed, the book's copy editor would place Wilbur among the experimentalists. Fortunately, she also noted that moving Wilbur into his more appropriate location would not substantially change the expected career peaks with respect to the four literary types.

years later than the conceptualist poets and novelists, novelists of both types create their best work more than 5 years later than the poets of both types, and that's after controlling for the life cycle effects. To sum up, genres and life cycles offer two independent sources of information about when literary geniuses offer their most acclaimed work to the world.

Observe that if Lermontov's *A Hero of Our Times* is viewed as a conceptual novel, as seems justifiable, then the youthful age of its creation becomes more understandable—just a little younger that when F. Scott Fitzgerald wrote *The Great Gatsby*.

Probably the most critical implication of this reanalysis of Galenson's data is that the expected career peak for novelists can actually be earlier than the expected peak for poets—*if* the former are conceptualists and the latter experimentalists. It is striking that conceptual novelists all produce their best work at ages prior to the 40s, with the minor exception of James Joyce, whose *Ulysses* appeared when he was 40. *Ulysses* remains a conceptual novel par excellence. In complete contrast, Robert Frost was a premiere experimental poet whose works embody the accumulation of literary and life wisdom. Hence, Frost's career peak is placed at 48 years of age. What did he create then? The poem "Stopping by Woods on a Snowy Evening." If you don't already know it, read to see what it means to be an experimental rather than conceptual poet.

Wise Old Philosophers, Historians, and Scholars Something is missing from our discussion thus far: What are the supremely maturity-favoring domains? Experimental novelists don't reach a peak later than 51 years of age, the age at which Thomas Hardy wrote *Tess of the d'Urbervilles*. Can't some domains demand even more experience, and thus cause the peak to appear even later?

Yes, indeed—such domains include much philosophy, history, and various forms of scholarship. To pick a perhaps extreme example, the German philosopher Immanuel Kant didn't write *Critique of Pure Reason* until age 57; two other works that also define his intellectual contributions, *Critique of Practical Reason* and *Critique of Judgment*, did not come out until he was 64 and 66 years old, respectively. That's pretty darn mature.

These contrasts across domains are easily accommodated by the equation that generated the curve in figure 8.1. The formula contains two parameters concerning the creative process, the ideation rate (how rapidly the creator comes up with new ideas) and the elaboration rate (how quickly those ideas can be transformed into finished products). These two rates largely depend on the domain (and whether the activity within the domain is conceptual or experimental). For example, the rates are faster for lyric poetry and pure mathematics, but slower for large fiction and philosophy. The two rates can also vary independently of each other. For instance, the elaboration rate for pure mathematics can be slower than that for lyric poetry, given how long it takes to put together a proof these days! Anyhow, these two cognitive rates influence the location of the curve's peak as well as the magnitude of the post-peak decline. The placement of the first, best, and last hit then follows suit.

One-Shots to Non-Stops

Thus far our focus has been on works of quality, ignoring the underlying quantity of creative output on which those works are ultimately founded. Simply put, we've been looking at hits, not attempts. Yet when we return to figure 8.1, it becomes manifest that the hits stand atop an underlying mound of misses represented by the typical age-productivity curve. Unsuccessful output usually appears before the first hit and after the last hit, and even between the first and last hit various misses will often be interspersed among any of the other hits, including the best hit. The career as a whole looks largely like a trajectory of randomly distributed hits and misses, with the misses most often outnumbering the hits. It's like an archer trying to hit a bull's-eye; most shots will be off target, and some by quite a wide margin, but the successes and failures will come and go like the heads and tails on a flipped coin.

Let's go back Thomas Edison for a concrete example. As I noted under Tip 6, his superlatively successful career was riddled with horrendous failures, such as the electric cars and concrete houses. Moreover, these misses were interspersed throughout his creative life. He successfully applied for his very first patent, when 21 years old, for an electrographic vote-recorder that nobody wanted, legislators preferring the customary roll call of ayes and nays. His very last patent was granted for a method for extracting rubber from plants. Because he was 80 years old at the time, he never learned that the process was not commercially viable, given that the plants were presumed to contain very little sap from which to produce rubber—a failure that echoed his earlier and more financially disastrous attempt to extract iron from low-quality ore. Edison had something of the alchemist about him, always trying to convert lead into gold. Between the first and last non-hit patents are interspersed other obscure patents, such as ones for fruit preservation, fiber and fabric waterproofing, stencil pens, electric safety lanterns, magnetic belting, expansible pulleys, and phonograph cabinets. Between ages 33 and 65, Edison took out 53 patents on his failed enterprise of mining and ore milling. Yet scattered among these wasted efforts are the patents for the big hits that made his name legendary in

America. Although Edison's investors had been assured "a minor invention every 10 days and a big thing every six months or so," he wasn't always able to deliver on the promise—not even on average.

Naturally, Edison hedged his bets somewhat by investing heavily in a diversified portfolio—approximately 1,500 patent applications for dozens of different types of inventions. So woe to the one-shot inventor who places all of their hopes on a single ingenious idea. Such one-shots are actually commonplace, the norm in fact, but the odds of any particular one enjoying success is minuscule. The only way to increase the odds is to become a non-stop inventor like Edison. Then one or more attempts will become a hit.

This continual inventive activity presumes that the inventor can leave his or her day job, like Edison did when he stopped making a living as a telegrapher after finally making enough money from his inventions. Some inventors work for industrial laboratories to the same end, such as the famous technological creativity displayed at Bell Labs. Since the labs were founded in 1925, its workers invented the transistor, the laser, and the charge-coupled device, developed information theory and radio astronomy, created the Unix operating system, and devised programming languages like C and C++, while accumulating eight Nobel Prizes along the way. Edison himself helped establish institutionalized invention when he founded his own "idea factory" at Menlo Park, New Jersey.

Lifetime Output and the Career Course Increasing the output of ideas has an interesting effect on the expected career trajectory, as represented schematically in figure 8.1. Given two persons operating in the same domain, but who vary only in total productivity, then the curve given in the figure will remain unchanged, except that the output rate must increase, making the peak higher without altering its location or the curve's true shape. For the typical case in the figure, the maximum appears around age 42, where the productivity equals four attempts per year. For a creator who's twice as prolific, the optimum age would remain at 42, but the rate maximize at eight attempts per year. By the same token, for a creator half as prolific as the one the figure represents, the result would be two attempts per year, but with the peak still in the same place. The reason that the curve's shape remains unchanged is simply that the ideation and elaboration rates would presumably stay the same, too, under the assumption that the creators work in the same domain.

Yet what are the consequences for the location of the first, best, and last hit? For the best hit, the answer is easy: because the middle career landmark tends to appear near the productive peak, its placement is unaltered. So the age of 40 plus or minus still functions as an all-around estimate. But the first and last hits are strongly affected. As productivity increases, they become further apart, meaning that the first hit comes at a younger age and the last hit comes at an older age. Conversely, as productivity decreases, they come closer together, converging on the best hit. In the extreme case, as already observed, the first, best, and last hit become exactly the same hit, just like the only child.

A Haydnesque Exemplification The Austrian composer Franz Joseph Haydn shows how far apart the first and the last hits can lie in a non-stop productive career. Here's a genius who composed hundreds of works, including over a 100 symphonies alone. He started producing compositions in his late teens and continued until his early 70s, when severe illness obstructed his work until he passed away at age 77. The first composition to become part of the classical repertoire is his Symphony no. 6, which he wrote when he was 29 years old. Like his other symphonies that became highly popular, it was later nicknamed *Le matin* ("morning") because the first movement's introduction clearly depicts a sunrise. The last composition to enter the repertoire was Haydn's oratorio *The Seasons*, which he composed at age 69. Hence, his first and last hits are separated by 40 years—with so many additional hits between that it's very difficult to identify the single best.

Nevertheless, the Haydn composition that became his most frequently performed and recorded—one that is included as a standard in numerous "music appreciation" courses—is his Symphony no. 94, popularly known as the "Surprise" Symphony. This Haydn created when he was 59 years old, an atypical age that falls closer to his last hit than to the first. No, that's no surprise, because all of the trajectories we're talking about here are statistical rather than deterministic. Variation around some average is to be expected. Furthermore, because Haydn was much more productive in later life than most composers, that would shift the expected age of the best hit toward the more advanced years. No spoilers, however: if you don't already know the basis for the nickname, you'll have to listen; just hearing the second movement will do.

Early to Late Starters

Under Tip 7 I observed that the very notion of "late bloomer" is fraught with confusions. I then stressed that the distinction between an "early" and "late" bloomer would be here defined in reference to the mean age of the first hit for a particular domain. A first hit that emerges earlier than the average indicates "early" and one that appears later than the average indicates "late." At the time, that definition made sense because it was conceived in terms of the 10-year rule that describes the amount of expertise acquisition necessary before a creator can make a major contribution to a given domain of creative achievement—which is just another way of saying the first hit. For example, when advocates of the 10-year rule count off the years, they begin at the onset of so-called deliberate practice and then end with the first hit. In the case of classical music, for instance, the year count might start at the first composition lesson and end with the first composition that has become a lasting feature of the repertoire.

Yet in light of what was said in the previous section, and after thinking deeply about the typical career trajectory given in figure 8.1, it's clear that two entirely different concepts about the creative career have been inappropriately conflated. That confusion happens because between the end of the preparatory period of expertise acquisition and the creation of the first hit lies another critical career benchmark—the actual onset of creative productivity. At this point in the life of the creative genius, products are generated, all right, but they are not yet counted as genuine successes. Neither are these works apprentice pieces or juvenilia, for by the time they are produced, the creator is in full command of the relevant domain-specific expertise. It's just that the probability of a success is so low, that even competent products can fall by the wayside. Even after acquiring the requisite expertise, many attempts are required before a hit becomes likely. To go to a metaphor I used above, an archer may have fully mastered the skill set involved in shooting an arrow at a target, yet the target is so distant and the surrounding conditions so capricious—such as the wind and various distractions—that luck still intrudes on getting a bull's-eye.

Consequently, to be really fair, and scientifically precise besides, expertise acquisition should not be measured all the way to the first hit, but rather should stop at the point when the creative genius starts producing expert work, ignoring whether that work becomes a success. Before that point, study and practice are key, as the genius is still mastering the rules and skills of the game; after that point, the laws of chance take over because creativity then becomes a strictly hit-or-miss affair. No player in professional baseball is a novice when it comes to hitting a pitched ball, given all of the batting practice and games played. It's just that effectively connecting the bat to a fastball, curve ball, slider, changeup—or whatever other tricky throw the opposing pitcher has mastered—is never a sure bet. So the accumulated expertise can only allow you to increase the statistical odds. Here the saying "practice makes perfect" is plain wrong. Practice just allows you to stay competitive in a game of chance.

Ty Cobb to this day claims the highest batting average in the history of Major League Baseball. Do you know the stat? A measly 0.366 over the course of a 24-season career! That means that only a little over one third of the time at bat, he got a hit. So almost two-thirds of the attempts resulted in misses, including strikeouts, pop ups, and ground-outs. Although three times during his career he managed to exceed 0.400, he pulled that off only once in consecutive seasons, with a full decade separating the last two occasions (namely, 0.420 in 1911 and 0.409 in 1912, but 0.401 not until 1922). In addition, if you ignore his rookie year, in which he batted only 0.24, his batting average slightly declined across his professional career (albeit a more accurate description is that his stats show a slight increase before declining). All this granted, each time Ty Cobb walked up to the plate, we should assume that he brought the same batting expertise he had the last time he batted. Sure, he might be off his game for some transient reason or another, but in the end Cobb remained Cobb.

Starters versus Bloomers Hence, to avoid confusion, I now introduce the distinction between early and late *starters*, where the "start" refers to the onset of the creative career—when seeds are first planted before anything starts to bloom. For instance, Edison's first big hit was the phonograph, which he invented at age 30, but his career actually started with the invention that first received patent protection—the ill-fated electrographic vote-recorder—which was invented almost a decade earlier. To appreciate why this new developmental distinction is essential, look again at figure 8.1. The onset of the creative productivity curve is placed at age 20, even though the first hit appears much later. Yet this age at career onset is not set in stone. It can occur either earlier or later, such as age 15 or age 25. If the former, we have an early starter, and if the latter, a late starter. This contrast has obvious repercussions for the career trajectory. Taking two creators working in the same domain and displaying the same output rate, the early starter will have all three hits shifted to younger ages, whereas the late starter will have all three hits shifted to older ages. Accordingly, if their respective careers are assessed in terms of *career* age rather than chronological age, the two creators will exhibit identical trajectories. This is why the mathematical model used to generate the curve seen in the graph is actually defined in terms of career age. The clock doesn't start ticking until the onset of the career.

To be more accurate, early and late starters cannot enjoy equivalent trajectories if the latter starts too late. For when that happens, the anticipated trajectory abuts against the terminating career landmark—death. That inescapable reality implies that the creative career gets prematurely truncated. Potential masterpieces never see the light of day because their would-be creators have already entered into eternal darkness.

A Super-Late Starter I introduced Anton Bruckner as a late bloomer under Tip 7, but as a symphonist we might better view him as a late starter instead. His whole career as a composer of late-Romantic symphonies was shifted so that his first hit didn't occur until he was 50 years old, a decade after he initiated his symphonic studies. That was his suitably nicknamed "Romantic" Symphony no. 4. So his best hit should have appeared about a decade later, or around age 60. And, in fact, his Symphony no. 7 had its premiere performance when Bruckner was already 60 years old, and thus this work can be justifiably considered his best hit. It was certainly his biggest success during his lifetime (and my personal favorite besides). Given the locations of the first and best hits, the last hit should emerge about a dozen years later, according to the scheme in figure 8.1, which would put his last hit about age 72. Oops, he died at that age! The composer was right in the midst of completing the last movement of his Symphony no. 9. Realizing that he might not live to finish the movement, Bruckner suggested that his choral hymn *Te Deum* be played as a replacement, which seems fitting given that this very religious man had already dedicated the work to God. If only his career onset been a few months earlier, he would not have left behind an imperfect symphony as his last hit. And if his symphonic career had started a decade earlier, would a Symphony no. 10 have become his last hit instead?

Batting Averages Low to High

The last source of systematic departures from the pattern depicted in figure 8.1 concerns the creator's lifetime hit rate. As I mentioned under Tip 6, creative geniuses can consistently vary in their lifetime ratio of hits to attempts. In fact, the hit rates among creative geniuses can differ considerably more than the career batting averages in professional baseball. Great pitching and fielding make it extremely difficult to do any better than Ty Cobb's 0.366, while competent club managing prevents hitters from getting worse than George McBride's 0.218, the all-time worst for any non-pitcher with sufficient at-bats to enter the record books (the necessary compensation being his top-notch fielding). By comparison, eminent creators do not have the same competitive constraints. Nothing really prevents a creator from hitting the bull's-eye on every shot (especially if the person is so unambitious as to place the target unusually close). Off hand, I can't think of any creative genius with a hit rate of 100%. Even Mozart's best ratio of hits to attempts is less than twice Cobb's ratio of hits to at-bats. Pure one-hit wonders with a single attempt are also hard to find. Indeed, those who get a hit on their very first attempt may spend the remainder of their lives trying to prove that they aren't one-hit wonders, each failed attempt ironically lowering their lifetime hit rate.

There are plenty more cases of creative persons whose hit rates approach even if not equal zero. If Pachelbel can really only be credited with a single hit, then his ratio of hits to attempts must be smaller than 1:500, or a hit rate less than 0.2%. Wanna-be creative geniuses can definitely reach a zero percentage. The Scottish poet William McGonagall has gone down in history as the absolute worst poet in the English language; out of roughly 200 poems, not one is even considered good, and the worst are notoriously bad. The American filmmaker Ed Wood has been called the worst director of all time; not one hit out of many (low-budget) attempts, plus his film *Plan 9 from Outer Space* has been proclaimed the worst movie ever. Both McGonagall and Wood had hit rates equal to absolute zero, which disqualifies them as creative. That granted, they both might represent an odd guise of genius. After all, both are so bad as to attract a cult following, their devoted followers demonstrating their appreciation by the frequency and intensity of laughter! McGonagall's terrible poems remain in print, and videos of Wood's worst remain available for purchase. Wood even became the title character in a Tim Burton film starring Johnny Depp. A lot of great filmmakers have never gotten such A-list cinematic treatment! Being bad can be good, provided that the bad is bad enough.

In any event, if we confine attention to those authentic creators with non-zero hit rates, then it's easy to figure out how this factor affects their trajectories. Holding other factors constant, such as domain, output, and career onset, the higher the hit rate, the earlier the age at which the first hit appears and the later the age at which the last hit appears, whereas the expected age for the best hit remains unaltered. The only complication concerns whether the hit rate stays relatively constant across a creator's mature years. Some have argued that the rate declines, others that it increases, and yet others that it remains stable. We do know that at the level of the individual genius, it can go any of these ways. Einstein, Edison, and Mascagni all suffered losses in their success rates, whereas Mozart and the American songwriters Cole Porter and Irving Berlin all enjoyed gains in their success rates. Yet these gains and losses probably average out across large numbers of creators, yielding a fairly stable rate across the course of the career. Anyway, the variation in the rates across creators immensely exceeds the fluctuations in the rates within single creators. That secure generalization is all that we need to specify what's typical rather than idiosyncratic. Exceptions will always be with us no matter what. They're like those chronic drinkers and chain smokers who still live to be 104.

Extra Points for Picking Your Death Day?

We began with figure 8.1, which describes the typical career trajectory for a creative genius. The description begins with the creative productivity curve, with its onset after the end of expertise acquisition and its termination by death. Between onset and termination are then located the first, best, and last hits, the first two somewhat closer than the last two. Systematic departures from this broad life pattern can then be explicated by the intrusion of four added factors: (1) the domain of creative achievement, (2) differences in total output, (3) the age at career onset, and (4) variation in hit rates. Because these four factors operate largely independently of each other, their effects can be freely combined, thereby yielding descriptions that are closely tailored to the peculiarities of particular types of creative genius. For example, highly productive early starter novelists with low hit rates will exhibit a career pattern rather distinct from late starter poets who are not very productive but have high hit rates. Superimposed on these systematic contrasts, moreover, is the operation of chance. This addition of the luck factor then generates profiles even more distinctive, even accounting for the nitty-gritty details of singular creative geniuses. The random departures from the baseline expectations largely define each creator's individuality. What is left to explain after that?

From time to time, as well, I've talked about how the closing event of any career—death—relates to the first, best, and last hits. For instance, domains that favor earlier first, best, and last hits are also more likely to feature lower life expectancies. One possible explanation was that such creators *can* die young because they are more likely to get high-impact work out before they die. However, as seen in figure 8.2, chemists and physicists also tend to have their hits at younger ages—especially the best and last for chemists—without any corresponding decrement in life span. Hence, other explanations are possible, such as potential hazards to longevity associated with some domains. For example, under Tip 2 we observed that poets have a far higher vulnerability to the symptoms of mental illness than holds for any other domain of creative genius. That surely will not help the poet's longevity. Most notably, poets are substantially more likely to commit suicide than are creators in other domains. That should do it!

Outliving Your Genius

Apart from domain, the other three factors are also involved. The involvement is most obvious in the case of the age at career onset: everything else equal, particularly the age at death, the early starters are more likely to realize their full creative potential than are the late starters. I previously used Bruckner to illustrate how creativity can be tragically cut off. The music world would have been richer had he completed his Symphony no. 9, and perhaps even added a 10th besides. Yet it is also tragic in a different way when the downturn in the curve in figure 8.1 reaches the zero point prior to the death date. The creator then has officially "run dry," spending the final years without producing further work. The Finnish composer Jean Sibelius lived until he was 91 years old, but essentially composed nothing in the last three decades of his long life—notwithstanding rumors that he was working on an eighth symphony. Sibelius eventually became sensitive when anybody asked him about his creative silence. An infamous case in American literature is J. D. Salinger, who also died at age 91. His career as a fiction writer lasted only about a decade, starting with his single best work, *The Catcher in the Rye*, when he was 32, and ending with his last major work *Franny and Zooey*, when he was 42. By his mid-40s all original writing was over, albeit the three youthful stories written in his 20s were published after his death. In the remaining 40-plus years of his life, Salinger evolved into a mysterious recluse.

Sibelius and Salinger prove that you don't have to worry about leaving great ideas inside your head when you die provided that your creativity is exhausted beforehand. This lack of unfinished business is useful to the extent that achieved eminence is strongly dependent on the total number of masterpieces that are left for posterity to evaluate and appreciate. As a consequence, the ultimate posthumous reputation of any creative genius might increase with life span. The longer the life, the longer the career, and the higher the likelihood that the creator has squeezed out the final masterwork. One problem with this deduction is that it achieved eminence is basically uncorrelated with life span! Perhaps careers like those of Sibelius and Salinger serve to undermine the final assessment of a creator's worth. How creative could they be if they run out of steam so early, with 30 to 40 years and nothing to show for it? Were they really one-idea creators who managed to stretch out that idea beyond its usefulness until nothing remained? The French poet Arthur Rimbaud outdid both Sibelius and Salinger. He ceased all writing at age 21, and spent his remaining 16 years as a merchant traveling all over the world. Does that make you think more or less highly of his poetry? Was he then a mere flash in the pan?

Your Genius Cut Short

Just as critically, perhaps something is operating at the other end—with those who die unusually young. Does a creative genius whose life is cut tragically short receive an extra boost toward fame? Some indirect evidence might suggest so. For instance, political leaders, including both European monarchs and US presidents, gain some eminence credits should they die in office by violent means. The death via multiple gunshots of King Gustavus Adolphus of Sweden in the Battle of Lützen, the execution by guillotine of King Louis XVI of France, and the assassination of President Abraham Lincoln at Ford's Theater illustrate the effect. Yet in the case of creative genius, something beyond pure grief or sympathy may be operative. When a precocious genius dies young, admirers can always speculate about the great works that would have ensued if they had only lived to a ripe old age. What amazing mathematics might have emerged from Évariste Galois had he not been killed at 20! What phenomenal music might have come from Juan Crisóstomo Arriaga ("the Spanish Mozart") had he not succumbed to some undiagnosed lung disease at age 19! What inventive romantic poetry would Thomas Chatterton have produced if he had not decided to end it all by drinking arsenic when he was 17 years old! Do we have an empirical investigation addressing this possibility?

Yes we do, although I know only one study, so the results can only be considered tentative. The sample used for this investigation consisted of the 301 geniuses that Terman's student, Catharine Cox, had examined for her doctoral dissertation—the one

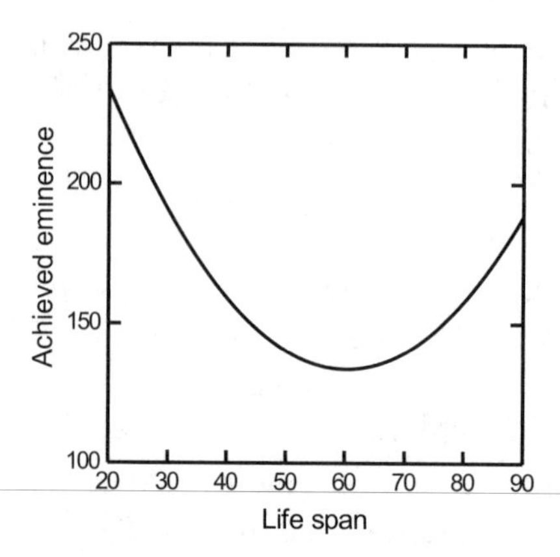

Figure 8.3
The relation between achieved eminence and life span for the 192 creative geniuses in the 1926 Cox sample of 301 geniuses.
Source: The figure uses an equation extracted from the statistics reported in Simonton 1976a, tables 1 and 2.

we discussed at length under Tip 1. Fully 192 of these geniuses were creators in the arts and sciences. Given their scores on achieved eminence and their calculated life spans, it's easy to estimate the relation between the two. The result is shown in figure 8.3.

Wow! The *least* distinguished creative geniuses are those who live to age 60, a rather ordinary longevity given that the mean for all 301 is around 66 years. These creators are surpassed in achieved eminence by those who live much longer than the norm—like Johann Wolfgang von Goethe who died at age 82, Lord Tennyson and Voltaire at age 83, Emanuel Swedenborg at 84, and Michelangelo at 88. Lastly, and most astonishing, the most eminent of all among the 192 are those who died much younger than the norm, such as Blaise Pascal at 39, Felix Mendelssohn at 38, Raphael at 37, Lord Byron at 36, and Wolfgang Amadeus Mozart at 35. Seemingly dramatic results!

Caveat Emptor!

Before you decide to take matters into your own hands to pursue the tragically young death route to creative genius, carefully contemplate the following three considerations. First, all of these early-death types were sufficiently precocious so as to produce multiple masterworks on which to stake their enduring eminence. If my reader is younger than 40, you must ask if you have done the same. Second, the curvilinear function shown in figure 8.3 has considerable scatter around the fitted curve, indicating that exceptions are very numerous and sometimes extreme. Chatterton was also in this sample, and died the youngest, so what did that get him? Why take the chance of doing much worse than Pascal and the rest? Finally, there's more to life than posthumous eminence, so why not enjoy it so long as you get well past age 60? As one poet warned, "Fame is a food that dead men eat." So please eat, drink, and be merry between each and every masterwork—and live a long and prosperous life into your 80s and beyond!

There! I think I've said enough to avoid any unreasonable lawsuits down the line!

Tip 9
Withdraw Alone to an Isolated Retreat / Social Network with Kindred Spirits!

The science of genius has largely been the pursuit of psychologists—scientists deeply fascinated with the individual human being. And of all human beings, creative geniuses count among the most fascinating groups they study. For the vast majority of us, the maximum we can hope for in the way of temporal immortality is a gravestone with an epitaph that remains legible to our kids and grandkids as long as they may live and care. In contrast, the geniuses who manage to create images, music, stories, ideas, proofs, theories, or inventions continue to leave an imprint for ages to come. Even when a creator's specific ideas become utterly obsolete over time, his or her impact will most often persist as an irrevocable residual. In that sense, they still survive the test of time.

Are Thomas Edison's incandescent light bulbs now getting progressively replaced by much more efficient forms of electrical lighting, such as light-emitting diodes? No problem. Who first figured out how to distribute inexpensive and usable electricity to entire neighborhoods and even cities, and thereby light up the world? Without that enormous and enduring achievement, where would LED or any other electrical device or appliance be anyway? Have those old-fashioned Edison cylinders become progressively replaced by 78-rpm disks and then long-playing records, only to see themselves pushed aside by 8-track tapes, cassettes, CDs, and downloaded mp3s? So what? Who first conceived the preposterous idea of providing an inventory of popular music available for purchase for whosoever had the cash—even if that music lover lived in the middle of nowhere? After a recording session on December 2, 1889, anyone in the world could listen to Johannes Brahms himself introducing and playing an excerpt from his very own Hungarian Dance no. 1. See the point? The contemporary details might have changed, but the initial revolutionary concept continues to shape our lives. The very notion that we might press an on-button, and then listen to our favorite music anywhere in the world, is easily traced back to Thomas Edison. Do you have to recharge your smart phone or tablet first? He dealt with that problem as well with his extensive research on secondary storage batteries: you can even buy an "Edison" battery today!

Still, not everybody believes that psychologists have any business studying creative genius. Once upon a time, I was a first-year graduate student in the social psychology program at Harvard's Department of Social Relations, which included sociology and cultural anthropology along with cognitive, personality, and social psychology. When I told one professor in the sociology wing of the program that I was interested in studying creative genius, he advised me that I had made a bad mistake: it was not a legitimate psychological subject, he said, because it was not a genuine psychological phenomenon! Any science of genius had to adopt a purely sociocultural perspective. Creativity was not a process that went on inside an individual's head, but rather it was the output of a social and cultural system. I was informed that the validity of this sociocultural perspective was demonstrated by the frequent occurrence of multiple discoveries and inventions. Multiples, as I call them for short, occur when two or more creators conceive the same idea independently of one another.

Indeed, some of the achievements attributed to Edison are more properly considered multiples. For example, the exact same year that Edison patented his successful incandescent lamp in the United States, Joseph Wilson Swan, an English physicist and chemist, patented his own incandescent lamp in Great Britain. Because their respective inventions were sufficiently close in conception, Edison and Swan were obliged to join forces to create the Edison & Swan United Electric Light Company, which eventually became known as Ediswan—and ultimately General Electric. Hence, a sociologist can easily argue that the incandescent lamp was not the unique creation of any singular genius. Instead, the concept was already "in the air" as part of the zeitgeist or "spirit of the times." If Edison or Swan had not been born, the technology would have emerged anyway, albeit with different surnames attached. The very fact that they both patented their lamps in 1879, making their inventions not just independent but simultaneous, offers particularly telling evidence for the operation of sociocultural determinism. In that year, that specific invention seems to have become inevitable, no matter who was around to get the nominal credit.

Many other examples of multiple discoveries or inventions can be given: the calculus by Isaac Newton and Gottfried Wilhelm Leibniz; the new planet Neptune by John Couch Adams and Urbain Le Verrier; the law of gases by Robert Boyle and Edme Mariotte; oxygen by Joseph Priestley and Carl Wilhelm Scheele; the periodic law of the elements by Alexandre-Émile

Béguyer de Chancourtois, John Newlands, Julius Lothar Meyer, and Dmitri Mendeleev; the theory of evolution by natural selection by Charles Darwin and Alfred Russel Wallace; the plague bacillus by Alexandre Yersin and Kitasato Shibasaburō; and the telephone by Alexander Graham Bell and Elisha Gray. The last multiple invention was as simultaneous as you can get, for the lawyers representing Bell and Gray showed up at the US Patent Office on the same day! Perhaps by some quirk, like which application made it to the top of the inbox, Bell's got processed first—hence the Bell telephone instead of the Gray.

So why did I persist in getting my degree in social psychology rather than in sociology? As I sensed then—and show later—multiples are not what they seem to be. They certainly cannot be used as empirical evidence for the impact of sociocultural determinism. The individual creative genius remains critical, and such individuals continue to have a psychology worth studying—as already proven under previous tips. At the same time, my PhD was received in *social* psychology, which implies that I also am inclined to view creativity in a larger context. But before I get to get to that side of the phenomenon, let me first discuss the argument for viewing the creative genius as a social isolate.

The Lone Genius Rides Again!

The image of the "lone genius" is right up there in popularity with the "mad genius." Here the expression can be used in two distinct senses. On the one hand, this pair of words might refer to the *personality* of geniuses, which might make them less sociable or extroverted. On the other hand, the expression could refer to the creative *productivity* of supreme geniuses, which requires that they adopt a more isolated life style. Naturally, these two aspects are intimately connected insofar as the personality inclination should make the productivity possible in the first place. That connection will become more obvious below.

Personality: Loners, Introverts, and Workaholics

The English poet William Wordsworth once famously described Isaac Newton as "a mind for ever / Voyaging through strange seas of Thought, alone." Nor is Newton alone in his aloneness. The inventory of eminent creators is crammed full of many other introverts if not outright loners: Frederic Chopin, Charles Darwin, Bob Dylan, Albert Einstein, Lady Gaga, Bill Gates, Jimi Hendrix, Elton John, Piet Mondrian, Beatrix Potter, Ayn Rand, J. K. Rowling, J. D. Salinger, Dr. Seuss (Theodor Seuss Geisel), Aleksandr Solzhenitsyn, Steven Spielberg, Nikola Tesla, Jay-Z (Shawn Corey Carter), and Mark Zuckerberg—just to give a random list of luminaries obtained through haphazard internet searches.

The problem is that such lists do not help us much, besides providing potential sources of anecdotes and quotations (such as the one above involving Wordsworth and Newton). If introversion represents the opposite end of a roughly normal distribution with extroversion at the other end, then even in the general population we should expect a significant proportion of introverts. What we really need is empirical evidence that creative genius shifts the distribution toward the introverted end. Under Tip 2 we saw that highly eminent artists were more likely to score high on traits that would not support good social mixing, such as being antisocial, cold, unempathetic, impersonal, egocentric, and tough-minded. Whether by personal choice or by the rejection of others, such individuals would not likely have others gravitate to them!

Schizothymic and Desurgent Folks R. B. Cattell, an eminent psychologist who devised a major personality questionnaire prior to the advent of the Big Five factors I've discussed under previous tips, provides more direct evidence. In particular, distinguished scientists—physicists, biologists, and psychologists—were found to score unusually high on schizothymia and desurgency. Sorry about dropping those terms into the mix: Cattell was notorious for coming up with neologisms to describe his newly devised personality factors. In plain English, the great scientists were inclined to be "withdrawn, skeptical, internally preoccupied, precise, and critical" (aka *schizothymic*), and to exhibit "introspectiveness, restraint, brooding, and solemnity of manner" (aka *desurgent*). Besides testing contemporary scientists, Cattell also estimated personality scores for deceased scientists for whom sufficient biographical information was available and came up with the same results.

One of Cattell's posthumously assessed subjects was Henry Cavendish, the English scientist best known for discovering hydrogen and for his ingenious experiments involving the accurate measurement of gravitational force. Cavendish later had the physics laboratory at Cambridge University named after him (a lab where 29 researchers have so far earned Nobel Prizes). How introverted was he? Cavendish rarely spoke to any male except with the fewest number of mumbled words, and never would speak with a woman at all, using written notes to communicate his orders to female servants. To avoid any human contact, he had a separate entrance to his house constructed for his exclusive use, and if any servant crossed his path, that poor soul would be fired right then and there! To escape unwanted social interactions on formal occasions, he would sometimes run away "squeaking like a bat." No doubt Cavendish represents the extreme endpoint on the introversion-extroversion scale, but he illustrates how far creative geniuses can sometimes go to avoid even the most minimal socializing. The life of the party they're not!

Cattell's basic results have been replicated using different methods and samples. The introverted disposition even shows up in recreational activities. For instance, Ravenna Helson's eminent female mathematicians were much more likely to engage in non-social leisure activities like listening to classical music records, reading the classics and other literature, and going for hikes. A similar pattern was seen in the 64 eminent scientists studied by Anne Roe. Rather than socializing, they vastly preferred "fishing, sailing, walking or some other individualistic activity." Furthermore, this loner inclination had its roots in childhood, when it was typical for the future scientist "to feel lonely and 'different' and to be shy and aloof from his classmates." This remark echoes what was said under Tip 7 concerning social adjustrnent problems too often faced by child prodigies. But that all 64 attained adulthood eminence, these problems didn't prevent the youths from realizing their potential.

Strong, Silent Types I hasten to add that the introversion of creative geniuses does not necessarily mandate that they must be shy or timid. Although they very well can be, as Cavendish shows, introverts can display social dominance, too. Because they are highly independent and autonomous, they are often more likely to be socially assertive than to conform to group pressures. Hence, when they enter into collaborative relationships, the collaborators will often have to assume more subordinate roles. As Einstein himself confessed, "I am a horse for a single harness, not cut out for tandem or teamwork, for well I know that in order to attain any definite goal, it is imperative that one person do the thinking and the commanding." This social dominance can sometimes lead an introvert to be mistaken for an extrovert, but only for the latter is any dominance tightly connected with broader sociability. Extroverts really like being with others, whereas introverts prefer being alone whenever possible. Or rather, as in Newton's case, introverts like being alone with their thoughts rather than having their minds bombarded by the thoughts of others.

Einstein's reference to a "definite goal" is key. For introverts, the goal almost always is to solve some problem, to achieve a preset task, to get the specific job done. Collaboration just serves as an unavoidable means to that end—like when Einstein was obliged to take on collaborators to do the math that was way beyond his expertise. For extroverts, the goal is more often taken to encompass group cohesion. An explicit task is not necessarily excluded, but its attainment cannot undermine the esprit de corps. Yet the latter consequence becomes definitely more likely when a group member, and especially the leader, adopts an "antisocial, cold, unempathetic, impersonal, egocentric, and tough-minded" interaction style.

Recognizing this distinction in social orientation leads us to the core contrast: creative geniuses are considerably more prone to be workaholics—in the sense that they seem "addicted" to their work. They have set high aspirations for themselves, and believe it's urgent to achieve them. Accordingly, Roe observed a standout feature of her 64 eminent scientists, namely, a "driving absorption in their work." Recall from Tip 1 what Catharine Cox similarly said about the superlative persistence seen in her historic geniuses. Why the struggle to work, work, work? Let's find out.

Productivity: Studios, Studies, and Laboratories

Go back to Tip 1. Given the choice to define IQ in terms of either tested intelligence or achieved eminence, I suggested that the latter enjoys superior long-term validity. A creative genius, in particular, attains eminence through enduring achievements in some culturally valued domain of creativity, whether in the arts or sciences. Those creative achievements then rather directly determine contemporary and posthumous reputation. In fact, the single most powerful predictor of eminence is the total number of such achievements, or, since Tip 5, what we have simply been calling "hits." On average, a one-hit wonder achieves less eminence than a two-hit creator, and the latter less eminence than a creative genius who generated has three distinct landmarks—the first, best, and last hit. Then there are the even greater geniuses who succeed in contributing multiple hits between the first and last hit. These are the folks with names like Thomas Edison, Galileo Galilei, Albert Einstein, Charles Darwin, Louis Pasteur, Fyodor Dostoyevsky, Vincent van Gogh, Oscar Niemeyer, Ingmar Bergman, Jean Sibelius, and Martha Graham—or you pick your own personal favorites. It just has to be a creator for whom you can identify more than a triad of creative achievements.

To be sure, the picture is complicated by the fact that creative domains differ in what can be considered a reasonable upper limit in the number of hits. Those who create in more modest forms can generate more hits than those who create in more ambitious forms—like songwriters versus opera composers or lyric poets versus novelists. It would definitely be most unfair to say a novelist is less eminent than a poet simply because the former created fewer novels than the latter created poems. Consequently, at the very minimum the eminence assessment must be rescaled according to the domain's baseline expectation. For example, each creator's total hit count might simply be divided by the maximum number of hits by creators in the same domain. Yet even after making that adjustment, the variation in the number of hits in any given domain is huge. Furthermore, the variation exhibits a very peculiar distribution.

The Lotka Law of Creative Productivity Unlike IQ scores, which roughly conform to the normal distribution described by the bell-shaped curve, the distribution of hits is so skewed that the one-hit wonders enormously outnumber the multiple-hit all-time creators. To fully appreciate the magnitude of this disparity, just examine figure 9.1.

Here the number of creative geniuses is a function of the number of hits each created. The curve stops at one hit because anything less would disqualify the creator as a genius. However, the termination at 10 hits was for convenience only. The hits could total a greater or lesser number according to the domain, but 10 hits serves well as an estimate. Now notice the oddity: although fully 100 creators can be credited with only a single hit, and thus they become one-hit wonders, only *one* creator pulled off the maximum of 10 hits, and thereby occupies the domain's creative peak alone. Just as dramatic is the curve connecting these two extremes. The decline from the left to the right is not gradual but rather precipitous. To give the actual numbers, after 100 creators produce 1 hit each, 25 produce 2 hits each, 11 yield 3, and then 6 → 4, 4 → 5, 3 → 6, 2 → 7 and 2 → 8, and, finally, 1 → 9 and 1 → 10. Those are the exact numbers that generated the graph. The number of creators adds up to 155, meaning that those with 3 hits or more represent only 19% of the entire sample of creators.

The formula that produced those numbers was derived from the Lotka law. Like all laws in the behavioral sciences, it's only an approximation, but it's still a reasonable first approximation.

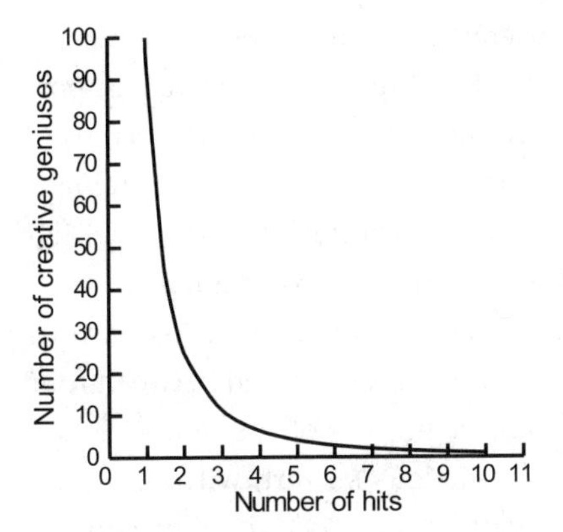

Figure 9.1
The number of creative geniuses who can claim a given number of hits over an entire career, where the former ranges from 1 to 100 and the latter ranges from 1 to 10.
Source: Based on Lotka's law, where $f(n) = 100/n^2$, $1 \leq n \leq 10$ (see Egghe 2005; Lotka 1926).

To get a better idea of how it might work in practice, we can use a concrete illustration drawn from the composers who have contributed lasting works to the operatic repertoire. For this purpose, I'll use statistics drawn from an official guide to the works performed at the New York Metropolitan, one of the premiere opera houses in the world. According to this guide, just 150 works constitute the "great operas" in the repertoire. These 150 operas were the work of 72 composers, who each produced an average of little more than 2 operas each. Yet the actual output is far from egalitarian. More than two-thirds of those composers, or 49, contributed just one opera each, making them the one-hit opera wonders. These 49 include Luigi Cherubini, Léo Delibes, Friedrich von Flotow, Alberto Ginastera, Mikhail Glinka, Scott

Joplin, Franz Lehár, Thea Musgrave, and many less known composers. On the other side of the distribution, solely two composers created 10 or more operas, namely, 10 by Richard Wagner and 15 by Giuseppe Verdi. Between these endpoints remain the 21 composers who produced the final 76 operas, which amounts to an average of less than four works each. In fact, the majority of these middling composers created just two or three operas. The very small minority who created between four and nine operas demarcates the beginning of the right-side tail of the distribution. Wolfgang Amadeus Mozart, with his seven great operas, falls in this secondary group. All in all, becoming an acclaimed creative genius in the opera house looks rather difficult. The two most successful, Verdi and Wagner, represent only about 3% of the whole, yet they created 17% of the standard repertoire!

The same holds for any creative domain worth its salt. Producing enough hits to put oneself at the very tip of the right-hand tail is by no means easy. As we learned under Tip 6, success often depends on having one failure or more. Even in the case of Mozart, his seven operatic hits represent only about a third of his total attempts. Cherubini's hit rate in opera was only about 3%, about a tenth as small as Mozart's. Moreover, a tremendous amount of effort lies behind every product no matter whether success or failure. Also, as I noted under Tip 6, the only route to a perfect product is an imperfect process—most often styled trial and error. Sometimes the imperfect process will yield one or more imperfect products before an acceptably perfect product emerges. Fortunately for the creator, the earlier versions are most often forgotten, inducing the misleading impression that the final result emerged like Minerva's mythological birth from the head of Zeus.

All That Work So Oft for Naught! To illustrate, consider the opera *Fidelio* by Ludwig van Beethoven. Even though this represents the composer's sole contribution to the operatic repertoire, it cannot count as his sole attempt. Beethoven's efforts began when he was given the libretto to an entirely different opera. Although he composed some music for that abortive effort, he dropped the idea entirely when he got the libretto for an early version of *Fidelio*. The composer was sufficiently resourceful to salvage two pieces from the earlier attempt to rework into an aria and a duet for the new attempt. The first version of the opera was performed about a year later, but was not successful. A new librettist helped shorten the opera from three acts to two, with corresponding revisions in the music, but the result was still highly flawed. Yet a third librettist worked on it, and Beethoven continued to do the same, until the final version premiered—the one that remains a regular on the opera stage more than two centuries later. All told, composition took more than a decade, and the opera even went through a name change after starting as *Leonore* (the heroine's real name) rather than *Fidelo* (the alias Leonore adopts when disguised as a male to save her husband from execution). As surviving testimony to Beethoven's trial-and-error efforts, the composer left behind three versions of the overture to *Leonore* and one final version for *Fidelio* in which he had started all over again from scratch. Ironically, the best of the three *Leonore* overtures (no. 3) had to be left on the cutting room floor because it was too good, seriously upstaging the opening scene that follows, as well as encapsulating the entire drama like one gigantic spoiler. Yet not all was lost, because that overture has now entered the standard orchestral repertoire as a stand-alone piece loved even by those concertgoers who hate opera. Such are the vagaries of creative genius!

The upshot is this: behind the scenes, creators are putting in an awesome amount of solitary effort just to produce a single hit on which they'll try to hang their future reputation. This means hours upon hours in the studio, study, or laboratory with perhaps nothing whatsoever to show for it. This hidden work most often presupposes a willingness to spend many hours alone, with a pen, pencil, brush, or keyboard ready at the fingertips, to record some new idea, or revise some old inspiration. Is anyone else besides an introvert fit for this kind of grueling work? What Roe said of her 64 eminent scientists can easily be paraphrased to apply to opera composers, songwriters, poets, novelists, or another domain in which creative genius operates: "Each works hard and devotedly at his laboratory, often seven days a week. He says his work is his life, and has few recreations. … [These scientists] have worked long hours for many years, frequently with no vacations to speak of, because they would rather be doing their work than anything else." At least the quotation suggests that there's still a potential payoff. Creative geniuses enjoy their work. Better yet, every once and awhile, an idea emerges and evolves into something that the creators can be really proud of, something that will enable them to leave a permanent mark on the domain—and perhaps even something worthy of engraving on their tombstones.

But Golden Ages of Creative Genius!

As we learned under Tip 3, Thomas Edison had a great love for Thomas Gray's "Elegy Written in a Country Churchyard," one of the best-known poems in the English language. As a young man, Edison memorized, and would frequently recite, the oft-quoted stanza:

The boast of heraldry, the pomp of pow'r,
And all that beauty, all that wealth e'er gave,
Awaits alike th' inevitable hour.
The paths of glory lead but to the grave.

You don't have to be a poetry aficionado to recognize the last line, or to recognize many other memorable inventions that have entered everyday language though the poem. Examples include "far from the madding crowd," "celestial fire," "the unlettered muse," "some mute inglorious Milton," and "kindred spirit," the last showing up in the title of this, the ninth tip! The poem's long-standing aesthetic impact is proven by the fact that it has been translated hundreds of times into more than three-dozen languages. Yet like Beethoven's *Fidelio*, Gray did not write the Elegy in one go, in some flash of infallible inspiration. Instead, he struggled through two or more versions over a period of about eight years. The poem was definitely completed in solitude—the introverted poet had by then retreated to a small village to engage in intense literary studies—and it specifically narrates a solitary meditation on mortality. At the superficial level, the poem's composition seems like an archetype of lone geniuses in action, contemplating their own death.

At a deeper level, though, this impression is completely wrong. Gray and his Elegy are so profoundly embedded in a larger context that a whole chapter could be written on the social psychology of the poem. To start with, the initial inspiration for it was the tragic loss of a beloved friend, and then its completion was likely ignited by the untimely death of a dear family member, who was buried in the churchyard featured in the poem (and where the poet himself was later interred). When he finally got something together that he was willing to share (apparently the penultimate version), he sent it off to a younger friend, Horace

Walpole, who was later to become an important literary figure in his own right. Walpole actually voiced a much more positive view of the poem than the poet originally expressed; thinking it would make a big splash, Walpole quickly circulated the creation among his literary friends. After Gray added the concluding epitaph, Robert Dodsley, an older contemporary who was a notable poet, playwright, and bookseller, formally published the Elegy in 1751. Beyond such professional contacts, this poem is enmeshed in contemporary literary developments, like the advent of the graveyard school, as represented by Robert Blair's "The Grave," published in the mid-1740s, right about the time Gray began to write the Elegy. In addition, the basic structure of Gray's poem—the "heroic quatrains" in iambic pentameter and *abab* rhyme scheme—had already been well established in earlier English poetry. The ideas contained in the poem may even betray some philosophical influences, such as the thinking of John Locke, the English empiricist active in the previous century. And who can miss Gray's shout-out to his great predecessor, the English poet John Milton? When facts such as these are combined with the influences that the Elegy exerted on subsequent poetry, particularly among the later Romantics, it would be easy to reconstruct the poem's date of composition from that contextual information alone. However introverted Gray might have been, the actual poem is encircled by a social network of causes and effects, poetic and otherwise.

So the time has come to yield the floor to the flip side of the ninth paradoxical tip. I'll start with a broad view of those moments in history when we might speak of the creative genius of whole civilizations. Then I'll narrow the focus a bit by scrutinizing the operation of creative domains. The two topics are irretrievably connected because the latter largely provides the engine behind the former.

Genius of Civilizations: Creative Times and Places

I mentioned earlier how I first learned about multiples from a sociologist, but a cultural anthropologist in the Social Relations Department might just as well have informed me. In fact, it was an eminent American cultural anthropologist, Alfred Kroeber, who, back in 1917, first drew attention to independent and simultaneous discovery and invention. He specifically used the phenomenon to argue for the superiority of the sociocultural milieu relative the lone genius. The latter, in his mind, were just epiphenomena with no real causal relevance. Among Kroeber's examples is the independent discovery of what are now known as Mendel's laws of genetics, first discovered by Gregor Mendel in his classic experiments on inheritance in peas. This discovery happened in 1865, but did not have a big impact at the time. And then suddenly, three scientists—the Dutch botanist Hugo de Vries, the German botanist Carl Correns, and the Austrian agronomist Erich von Tschermak—rediscovered the same laws in 1900. Because their discoveries occurred only a few weeks apart, they can be considered simultaneous as well as independent. Kroeber then argues that this event establishes a sociocultural determinism. In his own words, Mendelian heredity "was discovered in 1900 because it could have been discovered only then, and because it infallibly must have been discovered then."

Kroeber was using multiples in an explicit attack on Francis Galton's individualistic theory of genius. Nor was this the only anti-Galton argument he used. Much later, in 1944, Kroeber published *Configurations of Culture Growth*, which reports a series of empirical studies on the comings and goings of creative genius in world civilizations. In civilization after civilization he demonstrates that such genius is not randomly distributed across time and place but rather clusters rather dramatically into Golden Ages and perhaps subsequent Silver Ages, separated by periods that can only be called Dark Ages, when creative genius all but disappears from the scene—or rather starts appearing somewhere else on the globe. When Western civilization entered its own Dark Ages not long after "barbarians" overran the Roman Empire, new Golden Ages of creative genius were emerging first in China under the Tang dynasty and then in Islam under the Abbasid caliphate. The geographic centers of creative activity shift just as the intensity of creative activity fluctuates within a given center. Typically, each creative center will rise from obscurity, climb relatively quickly to a high point, and thereafter decline somewhat more gradually into mediocrity, at times surviving long enough to witness its hegemony replaced by some other center.

Two Historical Episodes I offer two illustrations here, one verbal and the other visual. Starting with the former, Greek philosophy first rose in Ionia, with Thales at Miletus, and spread to other Greek cities in the Mediterranean, as when Pythagoras of Samos moved to Croton in Italy. But after about 150 years, Greek thought converged on Athens as its center in the Golden Age. The climax started with Socrates, continued with his student Plato, and then recruited Plato's own student Aristotle, who emigrated from northern Greece. This Athenian triad defines the acme of ancient European philosophy—albeit the ideas of Socrates are only known through Plato's famed *Dialogues*. The impact of the latter works can hardly be overstated. As the modern English philosopher Alfred North Whitehead once noted, "The safest general characterization of the European philosophical tradition is that it consists of a series of footnotes to Plato." Most of the notable thinkers who were active toward the end of Greek philosophy, such as Plotinus in Egypt and Damascius back in Athens, were in fact Neoplatonists. Yet we must not overlook Aristotle's treatises on physics, biology, psychology, ethics, politics, poetics, rhetoric, logic, and metaphysics. These treatises exerted a profound influence on both Islamic and Christian philosophy in the Middle Ages. Indeed, in the *Summa Theologica*, which essentially defined Roman Catholic ideology, Thomas Aquinas cites Aristotle throughout, simply referring to him by the epithet "the Philosopher." No proper name necessary! Almost a full thousand years after Plato's Academy in Athens was closed down, the Italian Renaissance artist Raphael painted the masterpiece *The School of Athens* in the Vatican. Dominating the very center of the fresco are the impressive figures of Plato and Aristotle. In contrast, poor Socrates paid dearly for his failure to publish his thoughts, for he's shown lurking off to the left side!

Rather than offer a verbal description of a genius cluster, how about a visual representation? Figure 9.2 uses Kroeber's very own data to graph the number of scientific geniuses in the history of Islamic civilization from 700 to 1300 CE.

Although Islam began with Muhammad in the early seventh century, scientific activity did not start until the eighth. Even so, the first half of the eighth century produced no scientist of note, but in the latter half intellectual creativity, including scientific activity, got a big kick-start with the founding of the illustrious House of Wisdom in Bagdad by the Abbasid caliphs Harun al-Rashid and, especially, his son al-Ma'mun. The state-funded enterprise began with the translation of masterworks from

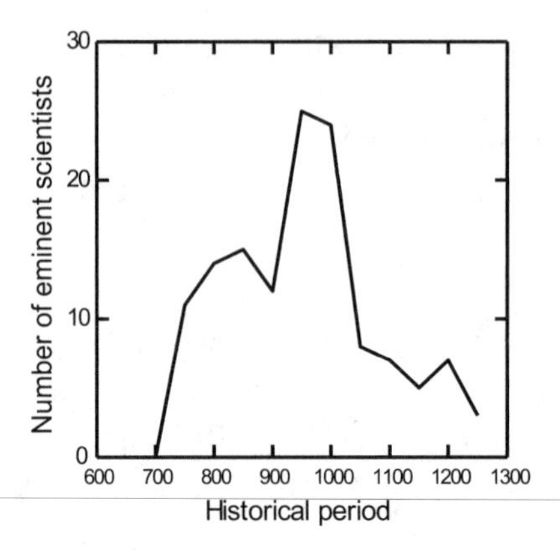

Figure 9.2
The number of eminent scientists active in Islamic civilization as a function of historical period, where the latter is broken down into consecutive half-centuries (700–749, 750–799, 800–849, etc.). The creators are tabulated into the half-century in which they attained their floruit, most often estimated as their 40th year.
Source: Based on data in Simonton 2016d, table 2 (cf. Simonton 2017b, figure 2, for a more fine-grained graph extending into the early 18th century).

other civilizations into Arabic. But from that impetus, emphasis quickly shifted to original work in mathematics, astronomy, alchemy (chemistry), medicine, biology, and geography, including cartography. The resulting Golden Age in science flourishes between 800 and 1050, during which every half-century produces a dozen or more great scientists, and the final century of that interval, from 950 to 1050, boasts at least two-dozen greats in each 50-year period.

A very partial list of even the most outstanding of these geniuses would name the mathematician al-Khwarizmi (the source for the eponym "algorithm"), the astronomer al-Farghani (after whom is named the lunar crater Alfraganus), the polymath Ibn Sina (known as Avicenna in the West, where his *Canon of Medicine* served as a standard medical text for centuries), the astronomer al-Battani (whose work was often used by Copernicus), the mathematician and physicist Ibn al-Haytham (aka Alhazen, the founder of optics), the astronomer Ibn Yunis (who also has a lunar crater named after him), and the polymath and polyglot al-Biruni (who mastered mathematics, astronomy, physics, geography, mineralogy, pharmacology, mineralogy, history, and Indology, learning Sanskrit for the purpose). After 1050, the decline is unusually precipitous in total count. That said, quantity of names is not perfectly indicative of the quality of those named: the last half of the 11th century features Omar Khayyám, one of the great polymaths of history—as if the Islamic Golden Age of science wished to end not with a whimper but with a final bang!

Origins of Cultural Configurations Given the ubiquity of configurations similar to the one depicted in figure 9.2, the obvious next question is where they come from. Like Kroeber argued, Galton's genetic determinism must be dismissed from the get-go. Even the most draconian eugenics program could not possibly have increased the gene pool sufficiently fast to explain the Golden Ages of Greek philosophy or Islamic science. The configurations just come and go way too fast. Hence, it's necessary to turn from nature to nurture for an explanation. The nurture account involves two types of environmental influences, external and internal.

External influences concern broad political, religious, economic, social, and cultural circumstances that impinge on the civilization, and which can deflect creativity up or down from baseline expectation. Actually, I already suggested this type of influence under Tip 3 when I mentioned Alphonse de Candolle's objection to the genetic determinism that Galton put forward in *Hereditary Genius*. Candolle had collected data showing that the uneven distribution of scientific genius among European nations could largely be attributed to contrasts in contextual variables that either supported or repressed scientific activity. Empirical research certainly endorses the impact of external conditions. For instance, in my Harvard doctoral dissertation I demonstrated that creative florescence in a civilization was a positive function of fragmentation into numerous independent states. The Golden Age of Greek philosophy certainly took place when Greek civilization displayed such fragmentation. Although warfare does not exert any broad impact, catastrophic instances definitely can. In the case of the Golden Age of Islamic science, it is significant that the House of Wisdom was utterly destroyed when invading Mongolian armies conquered and sacked Baghdad in 1258. After all, the lowest post-peak portion of the configuration shown in figure 9.2 appears in the latter half of the 13th century.

What about internal influences? Here we're talking about processes more intrinsic to the emergence of creative genius. Under Tip 3 I discussed how domain-specific mentors play a very important role in the development of creative genius. This was illustrated for Nobel laureates in figure 3.1. This developmental effect can be broadened to encompass all domain-specific role models who are potentially available in the prior generation. Just growing up in a milieu provided with an ample supply of active creative geniuses should facilitate creative development. The younger generation exposed to these exemplars would then prove more likely to become creative geniuses themselves.

Empirical studies of Western, Islamic, Chinese, and Japanese civilizations show that the number of creative geniuses in one generation tends to be a positive function of the number of creative geniuses in the previous generation. The creativity of each generation is thus building upon the creativity of its predecessors. More details about how this internal mechanism works follow below.

Genius in Domains, Fields, and Networks

The creativity researcher Mihaly Csikszentmihalyi may be a psychologist, but he still advanced the position that creativity is not a purely psychological phenomenon. Instead of creativity taking place within an individual's head, it is something that comes out of the interaction between the individual and a specific domain and field—the three components of his systems model. The *domain* is defined by a specific set of ideas, concepts, definitions, theorems, images, themes, materials, genres, styles, methods, techniques, goals, criteria, and so forth. Obviously, the ideational sets would differ greatly between artistic and scientific domains. Whereas on the one hand, Picasso's paintings could include nudes (however distorted beyond recognition), Einstein's scientific papers never did; on the other hand, Einstein's papers almost always contained recognizable mathematical functions, such as differential equations, but Picasso's paintings never did. Even so, closely related domains will share some of their ideational content. The mathematical sciences rely heavily on the calculus, for example. If you don't see the standard symbols for integration, differentiation, or some other operation, then it may not represent a mathematical science. To be sure, Einstein's brief paper in which he derived what eventually became known as $E = mc^2$ contains no calculus whatsoever, but the derivation was based on a much larger paper published earlier that year, which did use calculus quite extensively.

In contrast, the *field* consists of all those creators who are active in the same domain. These individuals include both predecessors (such as mentors) and contemporaries (such as colleagues and rivals). What they all share is the expertise and capacity to make creative contributions to the domain and to evaluate the contributions made by others. Picasso belonged to the field of early 20th-century painters and sculptors, among them Georges Braque, Juan Gris, Henri Laurens, Fernand Léger, Henri Matisse, and Diego Rivera; Einstein belonged to the field of early 20th-century theoretical physicists, including Max Born, Niels Bohr, Peter Debye, Arthur Eddington, Paul Ehrenfest, Max Planck, and Arnold Sommerfeld. In the sciences, fellow creators in the field are heavily involved in peer evaluation, so that those members determine acceptance of submitted manuscripts and grant proposals, as well as decide who receives major awards, such as Nobel Prizes. In the arts, however, evaluations often come from persons who are incapable of making creative contributions themselves but nonetheless can judge good work when they see it. These are the patrons, art dealers, editors, impresarios, connoisseurs, critics, and consumers of novels, poetry, painting, sculpture, architecture, music, dance, film, and so on. These recipients of artistic expression indeed may have opinions that differ from the creators themselves. In the domain of cinema, for instance, it is rare for a blockbuster to win critical acclaim.

The latter complications aside, the field plays the principal part in deciding the additions and subtractions to a domain—for creative domains cannot remain static over time and still remain creative. Once the field of chemists reached a consensus that phlogiston theory no longer worked, the very concept of phlogiston disappeared forever from the domain. (You can still find out what it is by googling it!) Once the field of biology realized that "Nothing in biology makes sense except in the light of evolution" (to use Theodosius Dobzhansky's noteworthy proclamation), then evolutionary theory became a mandatory component of the domain for all biologists. Antoine Lavoisier changed the domain of chemistry, just as Charles Darwin altered the domain of biology—but only after their revolutionary ideas had passed muster within their respective fields.

Analogous transformations occur in artistic domains as well. Colin Martindale, who I discussed briefly under Tip 4, devoted most of his career to showing how literary domains change over time. (His initial focus was French and English poetry.) Although poets at a particular time and place usually work under the constraints of a given aesthetic style, they are constantly driven toward increased originality within that style. As each poet surpasses the originality of their predecessors, the bar is raised for subsequent poets. Because this growth cannot go on forever, the style completely breaks down after a few generations, requiring replacement by a new style, like a shift from baroque to neoclassical or from neoclassical to Romantic. When this shift takes place, the very contents of the domain of poetry transform, such as the favored vocabulary, preferred themes, acceptable structures, and permissible metaphors. In the early Romantic period, the English poet William Wordsworth wrote "The Leach-Gatherer." Yes, it centers on a poor old man who was obliged to make ends meet by collecting blood-sucking worms for medical use by walking barelegged into marshes! That very choice of a dark metaphor for the poet's life would have been completely off limits for a pristine neoclassicist like Alexander Pope for whom all images came in polished marble!

Creative Cycles By now it should be clear that the individual, domain, and field interact to generate a cyclic process. The individual creator extracts from the domain the fundamental resources for generating potentially creative products. The field then evaluates these submitted products to determine which are worthy of explicit inclusion within the domain. At that point the potential creative products become actual creative products, via the field's consensus. The domain thus changed with additions and subtractions, a new batch of individuals pulls out a modified set of ideas, which then provide the foundation for more products. Moreover, it must be evident that a primary means by which a given generation acquires the domain-specific knowledge and skills is through the major creators active in the prior generation—those who can serve as domain-specific teachers, mentors, and role models. Consequently, Csikszentmihalyi's three-part cycle offers the underlying basis for Kroeber's configurations of culture growth.

Now, as the cultural configuration ascends toward the peak, the number of contemporaries—peers active in the same domain—necessarily increases. That's what happens in a Golden Age. Such peaks do not represent a single phenomenal creative genius existing without contemporaries. In fact, despite occasional exceptions like Omar Khayyám mentioned above, quality is positively correlated with quantity: the more total creators in a given period, the greater the greatest creators in that same period tend to be, on average. What this implies is that the greatest creative geniuses of history have the biggest opportunity to form relationships with multiple creators active in the same field. But do they actually do so? Or do they take the route of the lone genius, and lock themselves away in a remote retreat, never to be known to the world?

Creative Networks The correct answer to the above questions is that the top-notch creators apparently take full advantage of the social opportunities available at the acme of the configuration. The resultant networking with "kindred spirits" is seen in both artistic and scientific geniuses. In the former case, one study examined 772 painters and sculptors in the Western artistic tradition from the Renaissance to the 20th century. The more eminent the artist, the greater the number of direct interconnections with other eminent artists. Besides mentors in the prior generation, these relationships could entail rivals, collaborators, associates, friends, and co-pupils, as well as apprentices in the following generation—so that the network links at least three generations.

Parallel results were found for 2,026 eminent scientists. Newton may seem like a typical lone genius, yet he participated in controversies, rivalries, and competitions with 5 notable contemporaries, established friendships with 7 greats, and accumulated 26 professional contacts with various eminent correspondents, colleagues, or other associates. Nor were these relationships trivial or peripheral. Newton's friendship with Edmund Halley was instrumental in the eventual emergence of the *Principia*, much like Gray's friendship with Walpole facilitated the publication of the Elegy. And Newton's priority dispute with Leibniz over who first invented the calculus became an international cause célèbre that shaped the course of the history of mathematics. However introverted Newton's inherent personality, professional relations of one kind or another could often override that disposition when his scientific impact was at stake.

In the end, the members of your field represent your core constituency, whether you're an artist or scientist. The poet W. H. Auden made this point a tad crudely:

> The ideal audience the poet imagines consists of the beautiful who go to bed with him, the powerful who invite him to dinner and tell him secrets of state, and his fellow-poets. The actual audience he gets consists of myopic schoolteachers, pimply young men who eat in cafeterias, and his fellow-poets. This means that, in fact, he writes for his fellow-poets.

Even after Einstein became an enormously popular icon of his era, recognizable to millions of non-scientists all over the world, he still endeavored to impress his fellow theoretical physicists, not the average person on the street. His unified field theory did not fail because it was pop physics, but because it was bad physics.

Creative Decline Yet one crucial issue remains, as I suggest in the following questions: Why is the rise of the cultural configuration followed by the fall? Why can't the output of creative genius rise to ever-greater heights? Or, at the minimum, why shouldn't a civilization's creative florescence level off once the "carrying capacity" is reached, like the logistic curve describing population growth? Kroeber himself explained decline in terms of the "exhaustion" of the given cultural pattern. Each civilization begins with a set of possibilities that are used up in the process of becoming fully actualized by its creative geniuses. After the peak of the configuration, which represents the "classic" culmination of these possibilities, the remaining generations are left with increasingly more scraps or dregs, until further efforts become more pedantic or conforming or imitative than truly innovative. If creative geniuses can run out of ideas after their own career peaks, like Jean Sibelius and J. D. Salinger definitely did (as I discussed under Tip 8), why can't the civilizations of which they are a part do the same? Sounds pretty depressing, like the prognosis of the "end of science" briefly mentioned in the prologue. If this is a valid inference, then the only remaining curiosity is when we should expect the forthcoming Dark Ages to arrive! Do we need to make plans?

Happily, revivals can take place. I noted above how Martindale's empirical work indicates that poetic creativity in a given literary tradition might reboot by adopting a novel artistic style once the old one has already been run into the ground. Modern English poets stopped writing like Shakespeare long ago, yet there's plenty of great modern poetry to go around. At the level of the whole civilization, this cultural resuscitation often requires the intense infusion of ideas from alien civilizations. This infusion can result from extensive travels abroad by natives or substantial immigration from foreign cultures (recall the discussion under Tip 3 about the high creativity of immigrants). Such a cultural remix can renew a civilization's creative activity—as occurred in East Asia when Buddhism arrived from far-away India. The same booster shot can even launch an entirely new civilization from ground zero.

A stark example of the latter contingency is Bagdad's House of Wisdom, which I introduced above. In the late eighth century, the classic civilizations of Europe, the Middle East, and Northern India were already well past their prime. Yet a huge store of extant books in Greek, Persian, Syriac, and Sanskrit bore record of prior creative glories. The Abbasid caliphs then recruited not just Muslims, but also Christians and Jews, to translate these works into Arabic in order to create what was soon to become the largest library in the world—and thus ignite the "configuration of culture growth" graphed in figure 9.2. So amazingly thorough was this effort that many masterworks of antiquity are only known through their Arabic translations! Can creativity in modern civilization continue to thrive by similarly adopting an active incorporation of the best that other world cultures have to offer? If so, which cultures? How many? And how deep?

Genius and Zeitgeist Coexist!

Under Tip 6 we saw that even the most perfect creative products must emerge from imperfect creative processes or procedures that rely on some version of trial and error. This observation can be integrated with Csikszentmihalyi's systems model to yield a comprehensive theory of creative genius in a sociocultural context. The integration again starts with the domain containing the complete set of ideas. From this domain, each individual member of the field extracts a personal sample of domain-specific knowledge and techniques. I say "sample" because it is very rare if not impossible for most creators to fully master an entire domain. It would most often take a lifetime to do so, leaving no time left for creativity! Moreover, this restriction reflects the fact that during the acquisition of domain-specific expertise, individuals will be exposed to variable sources of information—different teachers, courses, textbooks, graduate programs, mentors, and so forth. In addition, because highly creative people display much more openness to experience, they will often add some idiosyncratic elements to their developing expertise. Recall how George Green taught himself a different calculus than he would have learned had he enrolled in calculus classes at any university in Britain. Furthermore, the acquired expertise may even include ideas well outside the domain proper. Remember the examples of Galileo and Gell-Mann cited under Tip 5. The net result is a highly personalized sample of the domain's ideational contents with putatively extraneous elements thrown in for better or worse.

In any case, at a certain point, the individual switches from an emphasis on creative development to an emphasis on creative productivity. As we saw under Tip 7, this transition most commonly occurs when the creator reaches the mid-20s. Then the ideas making up the creator's personal sample will become subjected to combinatorial processes or procedures. Why combinatorial? Because we have solid reasons to believe that all creative products represent combinatorial outcomes. To begin with, creative geniuses themselves have left introspective reports that stress their reliance on the capacity to freely generate ideational combinations. Thus, Einstein reported that "combinatorial play" constituted "the essential feature in productive thought." The French mathematician Henri Poincaré provided a more vivid description of this combinatorial play when he mentioned a creative episode in which "ideas rose in clouds; I felt them collide until pairs interlocked, so to speak, making a stable combination." The combinatorial nature of the products is also evident in the products themselves. For instance, the Canadian philosopher Paul Thagard has carefully analyzed 100 top discoveries and 100 top inventions, demonstrating that each and every one can be interpreted as a unique combination of various ideas. We also see the combinatorial status of creative products in the arts. For example, the computer programs that most successfully simulate artistic creativity all operate according to a combinatorial mechanism.

Critically, the ideational combinations are generated without knowing beforehand which will work and which not, and thereby arises the need to engage in trial and error, generation and test, or blind variation and selective retention—whatever is your favorite term. Because bad combinations must vastly outnumber the good ones, some selection must be introduced. Such selection operates at two main levels. First, the creators themselves offer an initial assessment of what they think might be worthy of presenting to the field. If the combination passes this personal assessment, then it enters into a potentially creative product. Second, the field itself evaluates the submitted product to determine if it contains any new ideas that deserve entrance into the domain—through publication, performance, exhibition, construction, or some other appropriate means. As I've already noted, the domain thus altered, subsequent creators acquiring expertise within the domain will necessarily work with modified personal samples, and the cycle continues.

At this juncture, it should have become manifest that the sociocultural context—the zeitgeist or "spirit of the times"—is located at two distinct places in this modified systems model. At the onset of the cycle, the domain largely defines what ideas will enter into the combinatorial hopper for those individuals who will create in that domain. I say "largely" rather than "entirely" because each individual extracts a different personal sample, a sample that may also bring in ideas outside the domain. At the end of the cycle, members of the field select whatever combinatorial products they believe, by collective consensus, deserve to be added to the domain. Between the onset and the end—intervening between the domain and the field—dwells the creative genius. After all, the combinatorial processes and procedures will usually operate within singular human minds. Even in collaborations and problem-solving groups, each collaborator or member will walk away with a set of ideas that will inevitably undergo further combinatorial play in moments of solitude, whether during a daytime hike in the woods or experiencing a nighttime reverie before the fireplace. A disposition toward introversion becomes very handy insofar as it permits these solitary moments to even take place. Extroverted individuals who prefer to remain completely immersed in social interactions from morning to night would have their own creativity accordingly truncated. Trial-and-error thinking consumes time, much alone time.

But it's now time to return to the phenomenon that inspired my whole interest in the genius-versus-zeitgeist debate.

Multiples Deconstructed

The above three-system conception of creativity has some startling implications regarding multiple discovery and invention. I said at the beginning of this tip that multiples are not what they seem. To see why, consider four key issues regarding the appearance of multiples in science and technology: autonomy, similarity, inevitability, and simultaneity.

Autonomy versus Dependence

Many so-called multiples are not truly independent and so shouldn't be counted at all. According to the model, a time gap exists between the individual's conception of an idea and its eventual incorporation into the domain, and thus become certified public knowledge. A lot can happen within that interval, especially if the earlier creator delays publication for any reason. For example, two Englishmen, Chester Moore Hall and John Dollard, have been credited (three decades apart) for independently inventing the achromatic lens. The 1728 date attributed to Hall is somewhat misleading because he didn't patent his invention at that time, but kept it a secret in order to create the first refracting telescope largely free of color distortion. (Newton had said that such a telescope was impossible; hence he invented a reflecting telescope.) Through a somewhat convoluted chain of events, Dollard found out about Hall's concept, and because he had already been working on that very problem, he was in the perfect position to reconstruct Hall's secret device. So in 1758 Dollard obtained the patent, not Hall. Even so, it would be stretching the definition of multiples beyond recognition to say that Dollard's invention was *autonomous* relative to Hall's. From Hall, Dollard learned it could be done.

In the above episode, Dollard was perfectly honest. He didn't even enforce his patent during his lifetime (albeit his son did so after dad died). But sometimes a less scrupulous individual will claim an autonomous discovery without proper acknowledgment of a direct influence. Discussions of the multiple discovery of oxygen by Priestly and Scheele sometimes added a third name, that of Lavoisier. Although Lavoisier deserves credit for recognizing oxygen as a separate element, and for coining the name as well, he failed to disclose that he already knew about the prior discoveries. Scheele had sent him a letter on that very point, and Priestly advised Lavoisier of his discovery when visiting him in his laboratory. Not good ethics!

Similarity versus Discrepancy

The "multiple" invention of the achromatic lens entailed almost exact duplicates—the combination of two lenses made of different shapes and materials that jointly canceled out the chromatic aberration produced by each. Yet most so-called multiples are very remote from representing genuine duplicates. If different creators operate on idiosyncratic personal samples drawn from the same domain, the odds are infinitely small that a blind combinatorial mechanism would generate absolutely identical combinations. For example, although both Newton and Leibniz can be credited with inventing the calculus, their two versions were radically different. Newton's version was so cumbersome, restrictive, and unintuitive that it inhibited the progress of abstract mathematics in England for more than a century. The version used in calculus classes all over the world today is that of Leibnitz, not Newton.

Many so-called multiples are so divergent that nobody recognized them as similar at the time. It was only much later, with the retrospective blurring of the details, that the act of lumping together overruled the splitting of hairs, thus unifying disparate concepts under a single generic concept. For instance, although steam turbines can be consolidated into a single generic "multiple," their separate designs are so contrasting that the very term *multiple* becomes meaningless.

Inevitability versus Coincidence

If we delete all putative multiples that were neither autonomously conceived nor truly identical, then a residual remains of genuine examples. A case in point is the Pelton water wheel, an invention that can also be credited to F. G. Hesse (even if the two conceptions had rather contrasting conceptual origins). Yet these bona fide multiples still differ regarding their *grade*, that is, the number of independent discoveries or inventions actually involved. Some multiples are grade 2 ("doublets"), others grade 3 ("triplets"), still others grade 4 ("quadruplets"), and so on. If multiples were generated by a deterministic process, as required by sociocultural determinism, then high-grade multiples should be far more common than low-grade multiples. Yet the absolute reverse holds. The frequency of a multiple grade declines very precipitously with the magnitude of the grade, rendering the doublets far more common than all of the other grades put together. Worse still, "singletons" that represent discoveries or inventions conceived only once outnumber all of the multiples of any grade whatsoever.

Tellingly, these empirically observed grade frequencies correspond very closely to what is known as the Poisson distribution. Why is that particular distribution important? It just so happens that the Poisson distribution is most descriptive of events that have such a low probability that any co-occurrence must be extremely rare. For instance, the Poisson distribution described the number of Prussian officers killed by a horse kick in a given year! Need I add that this empirical result throws any deterministic explanation, sociocultural or otherwise, immediately into the trashcan? Instead, the finding is far more supportive of the conclusion that the combinatorial mechanism is driven mostly by chance. This inference falls right into line with what we learned earlier about the place of randomness in creative productivity. Indeed, not only are hits a probabilistic consequence of the number of attempts, but a creator's participation in a multiple is a probabilistic consequence of the number of hits. The more prolific, the higher the odds of duplicated effort.

Simultaneity versus Rediscovery

But sociocultural determinists might protest, "What about all of the simultaneous multiples?" Shouldn't their appearance compel us to infer that at a specific moment in history, a particular discovery or invention becomes absolutely inevitable? Like Kroeber argued with respect to the supposed triplicate rediscovery of Mendel's laws? There are two big problems with this argument.

First, too many supposed multiples are far from simultaneous. For example, although one empirical inquiry found that a fifth of all multiples occurred within a single year, more than one-third required a time delay of a decade or more before all independent claims were complete. The rediscovery of Mendel's laws offers an obvious case given the 35-year delay between the monk's original discovery and the later rediscoveries. If the zeitgeist renders a particular discovery or invention absolutely inevitable at a specific moment in history, then where does this temporal slippage come from? How could Mendel be "ahead of his time" if the times are responsible for the discovery in the first place? Nor can it be argued that the zeitgeist varies by nation, so that Mendel's Austrian *ortgeist* was somehow more advanced that of the Netherlands (de Vries) or Germany (Correns). The other supposed co-rediscoverer, Erich von Tschermak, was also Austrian, and Tschermak's maternal grandfather had actually taught Mendel botany when Mendel was a student at the University of Vienna! That explanation just doesn't resolve the discrepancy.

Second, the degree of simultaneity is most likely indicative of a totally different phenomenon. Rather than indicate the causal convergence of mysterious sociocultural forces at a specific instant in history, simultaneity may reflect the rapidity by which a new idea is published and disseminated. As our systems model has it, some lapse of time may be required before the individual submits the ideational combination to the field for evaluation, and some more additional time before the novelty enters the domain sufficiently to become well known. Hence, as this evaluation and communication process becomes more efficient, simultaneity should increase regardless of the initial likelihood of the particular idea. In point of fact, the mean time interval between duplicates has decreased from around 86 years in the 16th century to about 2 years in the 20th century. Better yet, because more effective dissemination of knowledge more quickly slams the door on further multiples, the expected grades of multiples should decrease, and that's the case as well. With the advent of the internet, simultaneity has likely increased all the more and multiple grades correspondingly decreased. Nowadays, a big discovery or invention will be posted on some website almost instantaneously, at once terminating any opportunity for a multiple to occur.

Epitaph

Given the foregoing four points, multiples cannot possibly be taken as proof that creative genius represents a mere epiphenomenon. The sociocultural system is fundamentally incapable of supporting the hypothesized deterministic effects. Individual creativity still exists, notwithstanding the prominent role played by the domain and field. Contrary to what some sociologists and cultural anthropologists might have believed, sociocultural determinism is dead and buried! If Copernicus, Galileo, Descartes, Newton, or Einstein had died in their cribs, the world of physics would look very different today! The same can be said about any other domain of achievement had the creative geniuses who eventually shaped it come to that same fate.

Just as significant, even with all of the networking going on, ample room still remains for an introvert to prosper. So, please feel free to sneak off to your studio, study, or laboratory! You don't have to stay on Facebook all the time!

Epilogue

So how did you do on the *Genius Checklist*? Are you well on your way to creative genius? Or do you still have a huge amount of work ahead of you? Or, worse, has the train already left the station while you're madly trying to get a taxi? Perhaps it will help you a bit if we run briefly through the paradoxical tips one more time. And don't forget: these diverse considerations play off each other, so that a negative here may be ameliorated by a plus there!

Tip 1: Score at Least 140 on an IQ Test / Don't Even Bother Taking the Test!

Here you have two alternative routes to genius status: take a standardized IQ test and ace it with a score of 140 or more, or else make enduring contributions to whatever culturally valued domain of creative achievement you chose. The former path is much easier than the latter path, provided your general intelligence resides in the top 1% of the population. You can also become a bona fide genius at a very young age rather than wait until adulthood. As always, there's a catch. Although a high IQ can guarantee recognition as a genius according to one dictionary definition, it cannot possibly confer on you any claim to *creative* genius. If you want that prestigious qualifier, then even a high score on both creativity and IQ tests will not do the trick. The only remaining option is to abide by a second dictionary definition: just demonstrate "native intellectual power of an exalted type, such as is attributed to those who are esteemed greatest in any department of art, speculation, or practice; [or reveal an] instinctive and extraordinary capacity for imaginative creation, original thought, invention, or discovery." By this criterion, you will need very many years to pass the test, not just an hour or two. Even worse, it's conceivable that you won't live to find out your final score in the annals of history. Instead, you pass away as a neglected genius. An IQ test is hard to take, but the test of time can be outright cruel. But, as I said, it's your choice!

Tip 2: Go Stark Raving Mad / Become the Poster Child for Sanity!

The paradoxical advice of this tip is perhaps the trickiest of all nine. As I note, the mad-genius controversy boils over with contentious claims, many of them utterly unjustified, whether logically or empirically. First and foremost, nobody *must* exhibit debilitating mental illness to attain the highest levels of creative genius—too many supremely healthy brains among the greats for that to hold. Yet, at the same time, the clear presence of subclinical symptoms does not rule you out of the competition either, and certain symptoms, if sufficiently temperate in frequency and intensity, can actually enhance creative genius. Hence arises a sweet spot between pure normality and unadulterated insanity. Complicating matters all the more is that the specific location of this optimum on the health-illness continuum is heavily contingent on the specific domain of creative activity. This dependence goes beyond the contrast between scientific genius and artistic genius, involving prominent differences within separate sciences and arts. As if the ambiguities could not get any worse, general intelligence and other personal strengths likely have a critical role in regulating the more negative repercussions of psychopathology. Given all of the discovered intricacies, I would hope that your personal calculations err on the side of caution—toward mental health rather than illness! If you have to take your meds to get the right balance, then so be it!

Tip 3: Start Out as a Zygote with Super Genes / Carefully Pick Your Home and School!

The nature-nurture debate rivals the mad-genius issue in the less than civil behavior in which putatively objective psychological scientists often discuss the subject. In all likelihood, the decibel level of the shouting is proportional to the extremity of the stance taken on one side or the other. Although it's difficult to specify the precise middle ground between the extremes, any scientist open to the published data must maintain that genius is both born and made. On the genetic side, many individual differences associated with creative genius feature conspicuous contributions from the unique gifts granted at the moment of conception. These contributions concern both cognitive ability and personality traits, such as general intelligence and openness to experience. Yet on the experiential side, the creative development of genius is inevitably dependent on family and education. Even so, the impact of nurture can be a little quirky. To begin with, the non-shared environment seems to play a much bigger role than the shared environment, which explains why your siblings need not join you in your emergence as a genius. Don't feel guilty about that; it's just the way things go. More odd still, creative development is often nurtured by what seem to be very non-nurturing circumstances, such as traumatic events in childhood or adolescence. Fortunately, these diversifying experiences operate in a manner that seems to parallel those observed for mental illness. If you're going to become a physicist, avoid them, but if you want to become a poet, then seek them out. But, once more, be careful how much you ask for! You *can* have too much of a bad thing.

Tip 4: Be the Oldest Kid in Your Family / Make Sure You're Born Last!

We now turn to a more delicate subject: birth order, a non-shared environmental experience that can induce sibling rivalries of sufficient force to spoil family holiday get-togethers. Sitting around the dining table are three rather contrasting sibs—a firstborn physicist, a lastborn poet, and a middle-born athlete, or some such configuration—who still can't manage to get along after years of being on their separate life tracks. Happily, I think my discussion of this tip might effectively diffuse the tension. Too many diverse developmental factors are involved—such as gender distributions and age gaps, to start with, as well as the specific domain of achievement into which each sibling ventured—to permit simplistic I-am-better-than-you arguments. Besides, the effects of ordinal position remain small enough to permit many exceptions, even if the effects are notably superior to what can be inferred from a sibling's astrological sign. Accordingly, you probably should avoid using birth order to argue who's smarter or more open to experience. Whether those precautions are sufficient to avoid yelling matches about who was mom's or dad's favorite child, especially when the argument's fueled by one too many glasses of your favorite alcohol beverage, is a different problem altogether. Again, save this factoid for your eventual biographer who will have much more fun with it than you can.

Tip 5: Study Hard All Day and Night / Indulge Your Wide Interests, Hobbies, and Travels!

There are only so many hours in a day, and many of those hours are already committed to the essentials of life, like eating and sleeping. That 24-hour limit renders the dilemma underlying this tip particularly crucial. On the one hand, you can maximize the hours spent on deliberate practice in your chosen domain. Want to become a great physicist some day? Then read more physics texts, review more, do more problem sets, spend more time on your lab assignments, study more for the exams, and so forth. Avoid distractions: definitely *don't* play a musical instrument, read random fiction and nonfiction, engage in individual or team sports, go to the movies or attend concerts, try out novel cuisines, sign up for community service, cultivate interpersonal relationships, and so forth. On the other hand, creativity, including creative genius, is strongly linked to openness to experience. That means wide interests, hobbies, and other activities that may have no apparent connection whatsoever to mastering the expertise required for doing physics. How can these two divergent activities be reconciled? One way out of this predicament is to have one or more innate talents that enable you to accelerate expertise acquisition while still fostering an open mind. Another resolution is to pick a domain that doesn't require the full implementation of the 10-year rule, such as founding a new domain from scratch or revolutionizing an old one. Do whatever works best with respect to your talents and interests.

Tip 6: Impeccable Perfectionists Rule / More Failures Mean More Successes!

The difficulty presented here is that true geniuses most often seek perfection in their creative products, or at least a reasonable approximation to perfection. Presumably the absence of any glaring flaws may yield a masterwork on which you can hang your contemporary and posthumous reputation. Yet all of the processes and procedures needed to attain that goal are decidedly imperfect. No straight, narrow, and certain path will take you where you want to go. That unavoidable imperfection in means is encapsulated by the commonly used term "trial and error," where errors—the failed attempts—most often far outnumber the actual successes, or "hits." Even then, hundreds, even thousands of trials can produce not a single hit in any form. So the net outcome of extended trial and error may yet leave a long trail of failures, with genuine successes more or less randomly distributed between. Although I joke about the possibility of augmenting your hit rate, I provide absolutely no guidance on how you might pull that off. Perfectionists who concentrate their efforts on a small number of products are not necessarily better off than the mass producers who do no more than hold their noses to the grindstone and hope for the best. Hence, the optimal advice is to prepare yourself for failure after failure with the aspiration that a success or two will finally come your way. That's not a very encouraging suggestion, to be sure, but certainly the most realistic. Only if you give up the incessant quest toward perfection can you succeed each and every time! Yet what can such flawed achievements possibly do for your standing with posterity?

Tip 7: Turn Yourself into a Child Prodigy / Wait Until You Can Become a Late Bloomer!

At this point, it may seem that the science of genius is as dismal a science as economics is reputed to be. Becoming a creative genius seems just as hard as getting rich, if not way harder. Yet in timing your climb to Parnassus, the news seems more optimistic. At one extreme, you have the option of becoming a child prodigy, and thus getting a pre-dawn start on your ascent. Yet for various reasons, you may find yourself ill prepared for the strenuous hike, and eventually fall by the wayside, not coming anywhere close to your goal—ending up yet another prodigy who fell short of genius. At the other extreme, you can wait until you're really ready, even taking your dear sweet time in launching your preparations for the trek. The net result is the carefully planned and highly visible route undertaken at high noon, and thus pursue the strategy of the late bloomer. One drawback with the latter strategy is that it's normally much harder to acquire the needed domain-specific expertise once you've already entered adulthood. Combining intense study and practice with a demanding day job and progeny running around the house will not be easy, so it may require much more than a decade before you're already to go. That's why I recommended that you strive for a more average course, neither early nor late but rather normal progress toward your creative career. That seems a more achievable goal.

Tip 8: Do Your Best to Die Tragically Young / Just Live to a Ripe Old Age!

The previous tip focused on a small slice of life, namely the period between the onset of the acquisition of domain-specific expertise and the appearance of the first hit—the very first contribution to that chosen domain that endures. Yet a creative career typically involves much more than this initial phase. Besides the first hit, there's also the best single hit, the last hit, and death, when all creativity must come to a stop. Rendering the picture all the more complicated, failed attempts will often occur before the first hit and after the last hit, and even be distributed between the first and last hit, with the best hit falling among them. This assertion is particularly valid for creative geniuses who are highly productive. The precise location of the first, best, and last hits across the career is then a function of a host of varied factors, such as the creative domain, the magnitude of total lifetime output, the timing of the first attempt, and the creator's overall hit rate. Yet leaving these considerations aside, the most imperative influence is life span. No best or last hit can occur after death. Although that seems to imply that the longer-lived creative geniuses will attain higher achieved eminence, occasionally a genius will receive a boost from dying tragically young. That admitted, please don't do it: too risky!

Tip 9: Withdraw Alone to an Isolated Retreat / Social Network with Kindred Spirits!

The last pair of paradoxical tips grapples with another central dispute: the genius versus zeitgeist debate. Are outstanding creative achievements truly the result of individual creators offering their ideas to the world? Or are such achievements wholly the product of the larger sociocultural system, reducing all creative geniuses to epiphenomena? If the latter holds, then you just wasted your time reading the first eight tips, for they all assume that creativity is very largely (even if not entirely) a psychological phenomenon—something that actually does take place within a person's head. After reviewing all of the evidence on behalf of the lone genius, that very solitary genius is then embedded in the larger sociocultural context. That context includes the famed Golden Ages in which creative geniuses engage in rich networks with other creators active in the same domain. Along the way, we encounter multiple discovery and invention, something often taken as definite proof of a sociocultural determinism that leaves no place at all for the individual scientific genius. But we saw that this inference has no basis in either logic or fact. The bottom line remains: Please feel completely free to retreat into your study, studio, or laboratory to be alone in your thoughts, but don't forget to interact with your fellow great minds. If you all work well together, you can jointly create a cultural florescence most worthy of posterity's admiration.

One More Thing

Let me add the hope that the *Genius Checklist* can make some contribution, however small, to some creative florescence, whether in our time or in the remote future! My actual words may be long forgotten, but may a potent residual of these paradoxical tips somehow persist.

Endnotes

By choosing to note only the topics or phrases listed below (and keying them as appropriate to full citations in the list of references to follow), I emphasize original research on creative genius and related topics. This research is seldom accessible to the general public, except that tiny portion made available through open access journals. I do not cite biographical or historical information easily obtainable through a Google search.

Prologue

"After Einstein: Scientific Genius Is Extinct": Simonton 2013a.

"end of science": Horgan 1996.

$E = mQ^2$: http://www.gsjournal.net/old/physics/zeeper.pdf.

science of genius: Simonton 2012b.

self-help books: Gelb 2002, Patrick 2013, and Robledo 2016.

"Baby Einstein": http://www.kidsii.com/brands/baby-einstein/.

"basketball genius"... Curry: https://nbagenius.co/tag/stephen-curry/.

these investigations began: Galton 1865.

contemporary research: Simonton 2014d

genius ... made rather than born: Ericsson & Pool 2016; Howe 1999.

Tip 1

"20 Celebrities With Genius IQs": http://www.educatetoadvance.com/20-celebrities-with-genius-iqs/.

IQ genius threshold: *American Heritage Electronic Dictionary* 1992.

Alfred Binet and Theodore Simon [testing]: Binet & Simon 1905.

Stanford-Binet Intelligence Scale: Terman 1916.

top 1%: Terman 1925.

Genetic Studies of Genius: Terman 1925–1959.

still being studied today: Duggan & Friedman 2014.

grew up [Termites]: Terman & Oden 1959.

women were expected to become homemakers: Tomlinson-Keasey 1990.

women with IQs exceeding 180: Feldman 1984.

successful men ... unsuccessful men: Terman & Oden 1947.

Alvarez: https://en.wikipedia.org/wiki/Luis_Walter_Alvarez and Wohl 2007.

"One of the most brilliant": Wohl 2007, p. 968.

Shockley: https://en.wikipedia.org/wiki/William_Shockley and Isaacson 2014.

Watson ... Feynman [sub-Mensa IQs]: Ericsson & Pool 2016.

IQ scores for geniuses long, long deceased: http://iml.jou.ufl.edu/projects/Spring04/Artigas/facts.htm (all scores were taken from table 12A in Cox 1926; but the IQ for Emanuel Swedenborg was recorded incorrectly, and hence omitted here).

following true story [re Catharine Cox]: Robinson & Simonton 2014.

already published list [of eminent historical figures]: Cattell 1903.

"I am 4 years old ... I know the clock": Terman 1917, p. 210.

be able to give their gender ... their eyes: Binet & Simon 1908.

301-genius study: Cox 1926 (n.b.: several aspects of her research have been simplified in this summary because her work is extremely and impressively sophisticated).

IQ ranged between 153 and 164: Simonton 1976a.

John Stewart Mill ... self-taught: Cox 1926, pp. 707–709.

intelligence-eminence correlation [replications]: Simonton 2009.

Marilyn vos Savant ... recorded IQ: McFarlan 1989.

leaders ... lower IQs: Cox 1926; Simonton 1976a.

20 points lower: Simonton & Song 2009.

too much of a good thing: Antonakis, House, & Simonton 2017.

presidents of the United States [IQs]: Simonton 2006.

genius more likely ... assigned to great creators than to great leaders: Murray 1989; see also Ball 2014; Murray 2014; Sternberg & Bridges 2014; cf. Suedfeld 2014.

"High but not the highest intelligence": Cox 1926, p. 187.

"By natural ability... instinct": Galton 1869, p. 33.

grit: Duckworth 2016; cf. Grohman, Ivcevic, Silvia, & Kaufman 2017.

"Between the ages ... opera": Cox 1926, p. 593.

"whilst he was playing ... by way of horse": Barrington 1770.

"Native intellectual power ... discovery": *American Heritage Electronic Dictionary* 1992.

authentic genius ... pervasive impression that endures: Ginsburgh & Weyers 2014; Over 1982; Rosengren 1985; Simonton 1991c, 1998.

Tip 2

"mad-genius controversy": Kyaga 2015; Lombroso 1891.

mad genius ... myth: Schlesinger 2009.

mad-genius controversy: Becker 1978.

either/or position: Simonton 2000c.

DSM-5: American Psychiatric Association 2013.

psychometric research: Kaufman 2014.

lifetime prevalence ... 50%: https://psychcentral.com/blog/archives/2011/09/03/cdc-statistics-mental-illness-in-the-us/ (see Kessler, Chiu, Demler, Merikangas, & Walters 2005, for somewhat lower estimate).

Two investigations ... [into] the mad-genius debate: Ludwig 1995; Post 1994.

at-a-distance personality assessment: Song & Simonton 2007.

flourishing: Keyes & Haidt 2003.

Creativity ... not good therapy: Kaufman & Sexton 2006.

Ludwig's data: Ludwig 1995, from figure 7.8.

Post's percentages (rounded off to integers): Post 1994, from table 1.

paradigm [definition]: Kuhn 1970, pp. 10–11.

difference in paradigmatic guidance ... in ... social [and] natural sciences: Fanelli & Glänzel 2013; Simonton 2015a.

Psychopathology ... eminence ... contributions: Ko & Kim 2008.

"persons in professions ... forms": Ludwig 1998.

empirical studies ... similar evidence: Jamison 1989.

Positive Psychology ... body of research: Cassandro & Simonton 2002.

mad-genius paradox: Simonton 2014b.

Oxford Book of English Verse [study based on]: Martindale 1995.

every imaginable domain of creativity [elitism in]: Murray 2003.

257 German painters and artists: Götz & Götz 1979.

psychoticism traits: Eysenck 1995.

56 ... creative writers: Barron 1969.

"the upper 15 percent ... this test": Barron 1969, p. 72.

Other instances: Cattell & Drevdahl 1955; Rushton 1990.

CAS ... five criteria: Ludwig 1995, appendix E.

"the presence of psychological 'unease'": Ludwig 1995, p. 194.

correlation with creative achievement ... achieved eminence: Damian & Simonton 2015; Simonton 2008a.

Post ... recent follow-up: Simonton 2014c.

measures of achieved eminence: Murray 2003.

bias of biographers: Schlesinger 2009.

"Great Wits ... divide": Dryden 1681.

CAQ: Carson, Peterson, & Higgins 2005 (visual arts items from Appendix).

CAQ scores … cognitive disinhibition [and IQ]: Carson 2014.

IQ 180: http://www.eoht.info/page/IQ+tables.

"The only difference … I'm not mad": http://www.salvadordali.com/quotes/.

Tip 3

Hereditary Genius: Galton 1869.

P. D. Q. Bach: https://www.schickele.com/.

"I could add the names": Galton 1869, p. 210.

Alphonse collected data … findings: Candolle 1873.

his survey respondents included: Hilts 1975.

"the phrase 'nature and nurture'": Galton 1874, p. 12.

alliterative use of nature and nurture: Teigen 1984.

Making of a Scientist: Roe 1953.

64 eminent scientists: http://www.amphilsoc.org/mole/view?docId=ead/Mss.B.R621-ead.xml.

Many subsequent studies: Berry 1981; Chambers 1964; Eiduson 1962; Feist 1993; Moulin 1955; Zuckerman 1977.

"one-third … at all": Galton 1874, p. 176.

"fifteen elite schools" Zuckerman 1977, p. 83.

"thirteen elite universities" Zuckerman 1977, p. 89.

elite universities: see also Crane 1965; Ellis 1926; Helmreich et al. 1980; Kinnier, Metha, Buki, & Rawa 1994; Simonton 1992a.

"it is striking … laureates": Zuckerman 1977, p. 99–100.

[Regarding other distinguished mentors or masters]: See also Boring & Boring 1948; Simonton 1984a, 1992a, 1992b.

study of twins: Galton 1883.

degree of heritability: Bouchard 2004.

"gene for genius": Johnson & Bouchard 2014.

"Newton's ancestry … maternal line": Galton 1869, p. 220.

sans a stellar family pedigree: Lykken 1998.

Michelangelo … comparable to Newton: Murray 2003 (both ranked #1).

diversifying experiences: Damian & Simonton 2014.

"help weaken the constraints … socialization": Simonton 2000b, p. 152.

multicultural encounters: Godart, Maddux, Shipilov, & Galinsky, 2015; Leung, Maddux, Galinsky & Chiu 2008; Maddux, Adam, & Galinsky, 2010; Saad et al. 2013; see also Hellmanzik 2014; Levin & Stephan 1999; Simonton 1997b.

immigrants: Bowerman 1947; Goertzel, Goertzel, & Goertzel, 1978; Helson & Crutchfield 1970.

recipients of the Nobel Prize: Berry 1981.

"seem to have remarkably uneventful lives": Berry 1981, p. 389.

a thousand eminent 20th-century personalities: Simonton 1986.

gifted adolescents: Schaefer & Anastasi 1968.

scientific innovation: Sulloway 2014.

openness … creative genius: McCrae & Greenberg 2014.

Big Five personality scale [OCEAN]: Bouchard 2004.

imaginative … sophisticated [openness traits]: John 1990.

openness … cognitive disinhibition … psychopathology: Carson 2014.

foster children ... identical twins ... foster siblings: Plomin, DeFries, Knopik, & Neiderhiser 2016.

"regression toward the mean": Galton 1889.

Tip 4

shared and *non-shared* environments: Plomin & Daniels 1987.

shared environment ... little or no effect: Bouchard 2004.

"How many brothers ... older ... younger": Galton 1874, p. 201.

"Only sons ... middle": Galton 1874, p. 25.

"(1) that elder sons ... (3) ... eldest sons": Galton 1874, p. 26.

64 eminent scientists: Roe 1953; see also Chambers 1964; Eiduson 1962; Terry 1989; Helmreich et al. 1980; Helson & Crutchfield 1970; but see Feist 1993.

"5 are the oldest ... at age 2": Roe 1953, p. 71.

"most of those ... older brothers": Roe 1953, p. 72.

Female mathematicians: Helson 1980.

Mills College: Helson 1990.

"who were successful ... did not have brothers": Helson 1990, p. 49.

female psychologists: Simonton 2008b.

high-IQ children: Hollingworth 1926, 1942.

gender roles gradually becoming more equal: Simonton 2017a.

family size [control for in studies]: Clark & Rice 1982; Simonton 2008b; Terry 1989; West 1960.

non-shared environment [low explanatory power]: Turkheimer & Waldron 2000.

birth order [low explanatory power]: Damian & Roberts 2015a, 2015b.

peer relations... important: Harris 1995.

teenagers derailed from development of talents: Csikszentmihalyi, Rathunde & Whalen 1993.

classical composers: Schubert, Wagner, & Schubert 1977.

creative writers: Bliss 1970.

scientists... innovations: Sulloway 2014.

curvilinear U-shaped function: Ellis 1926.

Termites: Terman 1925.

"confluence theory": Zajonc 1976, 1983.

multiple problems confront use of this theory: e.g., Rodgers, Cleveland, van den Oord, & Rowe 2000; cf. Sulloway 2007.

more than 200,000 Norwegians: Bjerkedal, Kristensen, Skjeret, & Brevik 2007.

2.3 [IQ] points: Sulloway 2007.

"pompous, pedantic, and priggish": Kramer 2006, p. 25.

firstborn gets "dethroned": Adler 1938.

youngest... revolutionaries: Stewart 1977; Sulloway 2010.

birth order has minimal impact: Rohrer, Egloff, & Schmukle, 2015; see also Damian & Roberts 2015b.

about half as small: Damian & Roberts 2015a.

[Darwin's] evolutionary divergence: Sulloway 2010.

"7 species of": Sulloway 2010, p. 86.

"Like Darwin's Galápagos finches": Sulloway 2010, pp. 87–88.

eminent female psychologists: Simonton 2008b.

status quo politicians: Stewart 1977.

dangerous sports: Nisbett 1968.

stealing bases: Sulloway & Zweigenhaft 2010.

"firstborns are achievement oriented…laterborn[s]…unconventional": Sulloway 2010, p. 106.

openness…six facets: McCrae & Greenberg 2014.

Rousseau: McCrae & Greenberg 2014.

Tip 5

"the book nobody read": Gingerich 2005 (who proves the contrary).

scientific work…game of chess: Chase & Simon 1973; Simon & Chase 1973.

10-year rule: Ericsson 1996.

10,000-hour rule: Gladwell 2008.

deliberate practice: Ericsson & Pool 2016.

other domains of high achievement: Ericsson, Charness, Feltovich, & Hoffman 2006.

classical composers: Hayes 1989.

child prodigies: Simonton 2016c.

"while 12 percent…early period": Hayes 1989, p. 294.

peak performance…creative genius not excepted: Ericsson 2014.

"The Two Disciplines of Psychology": Cronbach 1957.

120 composers: Simonton 2016c; see also Simonton 1991b.

Talent…accelerate the acquisition: Simonton 2008a, 2014a.

Mendelssohn [as an eminence]: Murray 2003 (percentile calculated from p. 134).

Nobel laureates in the sciences ... artistic hobbies: Root-Bernstein et al. 2008.

"interests extend ... learning" http://www.nobelprize.org/nobel_prizes/physics/laureates/1969/gell-mann-bio.html.

Galileo ... interests in both literature and the visual arts: Simonton 2012b.

degrees of versatility: Cassandro 1998; Cassandro & Simonton 2010; Simonton 1976a; White 1931.

2,012 creative geniuses: Cassandro 1998.

versatile creators ... more productive ... higher degrees of achieved eminence: Simonton 1976a, 2000a; Sulloway 1996.

Simon's ["irrelevant"] publications: e.g., Simon 1954, 1955, 1973, 1974 (deliberately picking those single authored so that source of expertise is known).

Galileo and Leeuwenhoek [comparing antithetical forms of looking]: Simonton 2012b.

"Almost always ... obviously these ... replace them": Kuhn 1970, p. 90.

Tip 6

one-hit wonders: Kozbelt 2008.

conscientiousness ... scientific achievement: Grosul & Feist 2014; cf. Feist 1998, 2014.

painstaking ... practical [as conscientiousness descriptors]: John 1990.

artistic creators ... conscientiousness: Feist 1998.

chess genius: Campitelli, Gobet, & Bilalic 2014.

one [computer] program ... rediscovers laws in physics: Langley, Simon, Bradshaw & Zythow 1987.

imitation ... musical styles: Cope 2014.

divergent thinking ... sep-con articulation thinking [list of processes and procedures]: Carson 2014; Finke, Ward, & Smith 1992; Guilford 1967; Mednick 1962; Ness 2013; Newell & Simon 1972; Rothenberg 2015; Simonton & Damian 2013.

everyday thinking: Weisberg 2014.

Guernica ... sketches ... reconstruction of his thinking: Simonton 2007b.

Minotaurarchy: Damian & Simonton 2011; Weisberg 2004.

generate and test ... selective retention [alternative names for trial and error]: Bain 1855/1977; Campbell 1960; Nickles 2003; Popper 1963; Skinner 1981.

Edison: Simonton 2015b.

quality ... probabilistic consequence of quantity: Simonton 1997; Sinatra et al. 2016.

"The chances are ... the minor": Bennett 1980, p. 15.

"bear the unmistakable stamp": Evans 1974, p. 1747.

[sonnet] unevenness is born out ... research: Simonton 1989, 1990.

define creativity ... "non-obvious criterion": Simonton 2016a.

avoid self-imitation: Martindale 1990.

"It's a fake" ... "I often paint fakes": Koestler 1964, p. 82 (although this episode could be apocryphal, the fact persists that the artist did "paint fakes" numerous times).

quantity/quality relation: Simonton 2010; Sinatra et al. 2016.

satisficing: Simon 1956.

Mozart's mature years... between 60% and 70% [quality-quantity ratio]: Kozbelt 2005.

Tip 7

mathematical genius... first hit: Simonton 1991a.

Prodigies, productivity, and eminence: Simonton 1991b, 2016c.

Child Prodigies Play Favorites with Domains: Winner 2014 (although she gives ample examples of artistic prodigies, these more rarely become artistic geniuses).

classical composers... first hits: Simonton 2016c; cf. Simonton 1991b.

"My own idear... wind": Ashford 1919, p. 53.

social development fails to keep pace: Winner 2014.

autobiography: Wiener 1953.

Terence Tao... antithesis to the tragic tales: Kell & Lubinski 2014.

Crystallizing experiences... empirical inquiry: Walters & Gardner 1986.

study of illustrious American psychologists: Simonton 1992a.

physicists... first high-impact contribution: Simonton 1991a.

expected hit age... first high-impact contribution: Raskin 1936; Simonton 1991a, 1991b, 1992a; Zusne 1976.

Tip 8

Figure 8.1 depicts: the figure uses the equation $p(t) = c(e^{-at} - e^{-bt})$, where $c = 50$, $a = 0.04$, $b = 0.05$, and $t =$ the chronological age minus 20 to yield the corresponding career age (cf. Simonton 1991a).

120 most eminent composers: Simonton 2016c; cf. Kozbelt 2014.

"It is a pity … king": Ewen 1965, p. 863.

studying the correlation between age and creative productivity since 1835: Quetelet 1968.

my first paper on the question in 1975: Simonton 1975a.

Age and Achievement: Lehman 1953.

"most types of poetry … other than short stories": Lehman 1953, p. 325.

poets can die before they turn 40: Kaufman 2003; Simonton 1974; see also McKay & Kaufman 2014.

increased precocity … lifespan: McCann 2001; Simonton 1977.

cross-cultural and transhistorical universal: Simonton 1975a.

mean life spans of poets and novelists: Simonton 1974.

contrary creative life cycles: Galenson 2001, 2005; cf. Ginsburgh & Weyers 2005; Hellmanzik 2014.

encompass scientific genius: Jones, Reedy, & Weinberg 2014.

ideation rate and the elaboration rate: Simonton 1997a.

"a minor invention": Wirth 2008, p. 7.

unsuccessful output … trajectory: Simonton 1997a.

career age rather than chronological age: Simonton 1991a, 1997a, 2010.

Hit rate … stable: Simonton 1988a, 1997a.

Mozart … gains in success rates: Kozbelt 2005.

Cole Porter and Irving Berlin: Hass & Weisberg 2009.

eminence … uncorrelated with lifespan: Simonton 1976a, 1977, 1991b, 1992a.

political leaders ... eminence credits ... violent death: Simonton 1984b, 2001.

"Fame is a food": Henry Austin Dobson, https://allpoetry.com/Fame-Is-A-Food-That-Dead-Men-Eat.

Tip 9

multiple discovery and invention: Lamb & Easton 1984.

achievements ... considered multiples [Edison through Gray]: Simonton 2004, table 2.2.

"a mind for ever": http://www.bartleby.com/145/ww289.html.

schizothymia and desurgency: Cattell & Drevdahl 1955.

"withdrawn, skeptical ... critical" to "introspectiveness ... solemnity of manner" Cattell 1963, p. 121.

"squeaking like a bat": Cattell 1963, p. 121.

basic results ... replicated : Chambers 1964; Feist 1998.

female mathematicians: Helson 1971.

"fishing, sailing ... aloof from his classmates": Roe 1952, p. 22.

"I am a horse ... commanding": Sorokin 1963, p. 274.

creative genius ... workaholics: Helmreich, Spence, & Pred 1988; Matthews, Helmreich, Beane, & Lucker 1980.

"driving absorption in their work": Roe 1952, p. 25.

predictor of eminence ... hits: Simonton 1977, 1997a; see also Albert 1975.

statistics drawn from ... guide to works performed ... Metropolitan [Opera]: Freeman 1984; cf. Simonton 2014b.

"Each works hard ... anything else": Roe 1952, p. 25.

enjoy their work: Chambers 1964.

"The boast of heraldry ... to the grave": https://www.poemhunter.com/poem/elegy-written-in-a-country-churchyard/.

"was discovered in 1900 ... discovered then": Kroeber 1917, p. 199.

Configurations of Culture Growth: Kroeber 1944; see also Hellmanzik 2014; Murray 2003; Schneider 1937.

"the safest ... footnotes to Plato": Whitehead 1978/1929, p. 63.

External influences: Candolle 1873; see also Hellmanzik 2014; Murray 2003.

positive function of fragmentation: Simonton 1975b; see also Naroll et al. 1971; Simonton 1976b; cf. Simonton & Ting 2010.

warfare: Borowiecki 2014; Simonton 1980.

Western, Islamic, Chinese and Japanese civilizations: Simonton 1975b, 1988b, 1997b, 2017b; Simonton & Ting 2010.

systems model: Csikszentmihalyi 2014; cf. Sternberg & Bridges 2014.

blockbuster ... critical acclaim: Simonton 2011.

"Nothing in biology ... evolution": Dobzhansky 1973, p. 125.

style ... replacement by a new style: Martindale 1990.

quality positively correlated to quantity: Simonton 1988b; Sinatra et al. 2016.

772 painter and sculptors: Simonton 1984a; see also Hellmanzik 2014.

2,026 eminent scientists: Simonton 1992b.

"The ideal audience ... fellow poets": Auden 1948, p. 176.

"exhaustion" ... cultural pattern: Kroeber 1944; cf. Murray 2014; Sorokin 1937–1941.

revivals can take place: Simonton 1997b.

Western civilization ... Dark Ages: Simonton 2016b.

systems model ... theory of creative genius: Simonton 2010.

"combinatorial play": Hadamard 1945, p. 142.

"ideas rose in clouds ... combinations": Poincaré 1921, p. 387.

100 top discoveries and 100 inventions: Thagard 2012.

achromatic lens: Ogburn & Thomas 1922.

steam turbines: Constant 1978.

multiples differ regarding their *grade*: Merton 1961.

Poisson distribution ... low probability: Simonton 1979.

degree of simultaneity: Merton 1961.

simultaneity ... increased ... multiple grades ... decreased: Brannigan & Wanner 1983.

References

Adler, A. (1938). *Social Interest: A Challenge to Mankind* (J. Linton & R. Vaughan, trans.). London: Faber & Faber.

Albert, R. S. (1975). Toward a behavioral definition of genius. *American Psychologist, 30,* 140–151.

American Heritage Electronic Dictionary (3rd ed.). (1992). Boston: Houghton Mifflin.

American Psychiatric Association. (2013). *Diagnostic and Statistical Manual of Mental Disorders* (5th ed.). Arlington, VA: American Psychiatric Association Publishing.

Antonakis, J., House, R. J., & Simonton, D. K. (2017). Can super smart leaders suffer too much from a good thing? The curvilinear effect of intelligence on perceived leadership behavior. *Journal of Applied Psychology, 102,* 1003–1021.

Ashford, D. (1919). *The Young Visiters: Or, Mr. Salteena's Plan.* New York: Doran.

Auden, W. H. (1948). Squares and oblongs. In R. Arnheim, W. H. Auden, K. Shapiro, & D. A. Stauffer (Eds.), *Poets at Work: Essays Based on the Modern Poetry Collection at the Lockwood Memorial Library, University of Buffalo* (pp. 163–181). New York: Harcourt, Brace.

Bain, A. (1977). In D. N. Robinson (Ed.), *The Senses and the Intellect.* Washington, DC: University Publications of America. (Original work published 1855)

Ball, L. (2014). The genius in history: Historiographic explorations. In D. K. Simonton (Ed.), *The Wiley Handbook of Genius* (pp. 3–19). Oxford: Wiley.

Barrington, D. (1770). Account of a very remarkable musician. *Philosophical Transactions of the Royal Society of London, 60,* 54–64.

Barron, F. X. (1969). *Creative Cerson and Creative Process.* New York: Holt, Rinehart & Winston.

Becker, G. (1978). *The Mad Genius Controversy: A Study in the Sociology of Deviance.* Beverly Hills, CA: Sage Publications.

Bennet, W. (1980). Providing for posterity. *Harvard Magazine, 82,* 13–16.

Berry, C. (1981). The Nobel scientists and the origins of scientific achievement. *British Journal of Sociology, 32,* 381–391.

Binet, A., & Simon, T. (1905). Méthodes nouvelles pour le diagnostic du niveau intellectuel des anormaux. *L'Année Psychologique, 12,* 191–244.

Binet, A., & Simon, T. (1908). The development of intelligence in the child. In W. Dennis & M. W. Dennis (Eds.), *The Intellectually Gifted: An Overview* (pp. 13–16). New York: Grune & Stratton.

Bjerkedal, T., Kristensen, P., Skjeret, G. A., & Brevik, J. I. (2007). Intelligence test scores and birth order among young Norwegian men (conscripts) analyzed within and between families. *Intelligence, 35,* 503–514.

Bliss, W. D. (1970). Birth order of creative writers. *Journal of Individual Psychology, 26,* 200–202.

Boring, M. D., & Boring, E. G. (1948). Masters and pupils among the American psychologists. *American Journal of Psychology, 61,* 527–534.

Bouchard, T. J., Jr. (2004). Genetic influence on human psychological traits: A survey. *Current Directions in Psychological Science, 13,* 148–151.

Bowerman, W. G. (1947). *Studies in Genius*. New York: Philosophical Library.

Borowiecki, K. J. (2014). Artistic creativity and extreme events: The heterogeneous impact of war on composers' production. *Poetics, 47*, 83–105.

Brannigan, A., & Wanner, R. A. (1983). Multiple discoveries in science: A test of the communication theory. *Canadian Journal of Sociology, 8*, 135–151.

Campbell, D. T. (1960). Blind variation and selective retention in creative thought as in other knowledge processes. *Psychological Review, 67*, 380–400.

Campitelli, G., Gobet, F., & Bilalic, M. (2014). Cognitive processes and development of chess genius: An integrative approach. In D. K. Simonton (Ed.), *The Wiley Handbook of Genius* (pp. 350–374). Oxford: Wiley.

de Candolle, A. (1873). *Histoire des sciences et des savants depuis deux siècles*. Geneve: Georg.

Carson, S. H. (2014). Cognitive disinhibition, creativity, and psychopathology. In D. K. Simonton (Ed.), *The Wiley Handbook of Genius* (pp. 198–221). Oxford: Wiley.

Carson, S., Peterson, J. B., & Higgins, D. M. (2005). Reliability, validity, and factor structure of the Creative Achievement Questionnaire. *Creativity Research Journal, 17*, 37–50.

Cassandro, V. J. (1998). Explaining premature mortality across fields of creative endeavor. *Journal of Personality, 66*, 805–833.

Cassandro, V. J., & Simonton, D. K. (2002). Creativity and genius. In C. L. M. Keyes & J. Haidt (Eds.), *Flourishing: Positive Psychology and the Life Well-Lived* (pp. 163–183). Washington, DC: American Psychological Association.

Cassandro, V. J., & Simonton, D. K. (2010). Versatility, openness to experience, and topical diversity in creative products: An exploratory historiometric analysis of scientists, philosophers, and writers. *Journal of Creative Behavior, 44,* 1–18.

Cattell, J. M. (1903). A statistical study of eminent men. *Popular Science Monthly, 62,* 359–377.

Cattell, R. B. (1963). The personality and motivation of the researcher from measurements of contemporaries and from biography. In C. W. Taylor & F. Barron (Eds.), *Scientific Creativity: Its Recognition and Development* (pp. 119–131). New York: Wiley.

Cattell, R. B., & Drevdahl, J. E. (1955). A comparison of the personality profile (16 P. F.) of eminent researchers with that of eminent teachers and administrators, and of the general population. *British Journal of Psychology, 46,* 248–261.

Chambers, J. A. (1964). Relating personality and biographical factors to scientific creativity. *Psychological Monographs: General and Applied, 78* (7, Whole No. 584).

Chase, W. G., & Simon, H. A. (1973). Perception in chess. *Cognitive Psychology, 4,* 55–81.

Clark, R. D., & Rice, G. A. (1982). Family constellations and eminence: The birth orders of Nobel Prize winners. *Journal of Psychology, 110,* 281–287.

Constant, E. W., II. (1978). On the diversity of co-evolution of technological multiples: Steam turbines and Pelton water wheels. *Social Studies of Science, 8,* 183–210.

Cope, D. (2014). Virtual genius. In D. K. Simonton (Ed.), *The Wiley Handbook of Genius* (pp. 166–182). Oxford: Wiley.

Cox, C. (1926). *The Early Mental Traits of Three Hundred Geniuses.* Stanford, CA: Stanford University Press.

Crane, D. (1965). Scientists at major and minor universities: A study of productivity and recognition. *American Sociological Review, 30,* 699–714.

Cronbach, L. J. (1957). The two disciplines of scientific psychology. *American Psychologist, 12*, 671–684.

Csikszentmihalyi, M. (2014). The systems model of creativity and its applications. In D. K. Simonton (Ed.), *The Wiley Handbook of Genius* (pp. 533–545). Oxford: Wiley.

Csikszentmihalyi, M., Rathunde, K., & Whalen, S. (1993). *Talented Teenagers: The Roots of Success and Failure*. Cambridge: Cambridge University Press.

Damian, R. I., & Roberts, B. W. (2015a). The associations of birth order with personality and intelligence in a representative sample of U.S. high school students. *Journal of Research in Personality, 58*, 96–105.

Damian, R. I., & Roberts, B. W. (2015b). Settling the debate on birth order and personality. *Proceedings of the National Academy of Sciences of the United States of America, 112*, 14119–14120.

Damian, R. I., & Simonton, D. K. (2011). From past to future art: The creative impact of Picasso's 1935 *Minotauromachy* on his 1937 *Guernica. Psychology of Aesthetics, Creativity, and the Arts, 5*, 360–369.

Damian, R. I., & Simonton, D. K. (2014). Diversifying experiences in the development of genius and their impact on creative cognition. In D. K. Simonton (Ed.), *The Wiley Handbook of Genius* (pp. 375–393). Oxford: Wiley.

Damian, R. I., & Simonton, D. K. (2015). Psychopathology, adversity, and creativity: Diversifying experiences in the development of eminent African Americans. *Journal of Personality and Social Psychology, 108*, 623–636.

Dobzhansky, T. (1973). Nothing in biology makes sense except in the light of evolution. *American Biology Teacher, 35*, 125–129.

Dryden, J. (1681). *Absalom and Achitophel: A Poem*. London: Davis.

Duckworth, A. (2016). *Grit: The Power of Passion and Perseverance*. New York: Scribner.

Duggan, K. A., & Friedman, H. S. (2014). Lifetime biopsychosocial trajectories of the Terman gifted children: Health, well-being, and longevity. In D. K. Simonton (Ed.), *The Wiley Handbook of Genius* (pp. 488–507). Oxford: Wiley.

Egghe, L. (2005). *Power Laws in the Information Production Process: Lotkaian Informetrics*. Oxford, UK: Elsevier.

Eiduson, B. T. (1962). *Scientists: Their Psychological World*. New York: Basic Books.

Ellis, H. (1926). *A Study of British Genius* (rev. ed.). Boston: Houghton Mifflin.

Ericsson, K. A. (1996). The acquisition of expert performance: An introduction to some of the issues. In K. A. Ericsson (Ed.), *The Road to Expert Performance: Empirical Evidence from the Arts and Sciences, Sports, and Games* (pp. 1–50). Mahwah, NJ: Erlbaum.

Ericsson, K. A. (2014). Creative genius: A view from the expert-performance approach. In D. K. Simonton (Ed.), *The Wiley Handbook of Genius* (pp. 321–349). Oxford: Wiley.

Ericsson, K. A., Charness, N., Feltovich, P. J., & Hoffman, R. R. (Eds.). (2006). *The Cambridge Handbook of Expertise and Expert Performance*. New York: Cambridge University Press.

Ericsson, A., & Pool, R. (2016). *Peak: Secrets from the New Science of Expertise*. Boston: Houghton Mifflin Harcourt.

Evans, G. B. (Ed.). (1974). *The Riverside Shakespeare*. Boston: Houghton Mifflin.

Ewen, D. (1965). *The Complete Book of Classical Music*. Englewood Cliffs, NJ: Prentice-Hall.

Eysenck, H. J. (1995). *Genius: The Natural History of Creativity*. Cambridge: Cambridge University Press.

Fanelli, D., & Glänzel, W. (2013). Bibliometric evidence for a hierarchy of the sciences. *PLoS One*, *8*(6), e66938. doi:10.1371/journal.pone.0066938.

Feist, G. J. (1993). A structural model of scientific eminence. *Psychological Science, 4,* 366–371.

Feist, G. J. (1998). A meta-analysis of personality in scientific and artistic creativity. *Personality and Social Psychology Review, 2,* 290–309.

Feist, G. J. (2014). Psychometric studies of scientific talent and eminence. In D. K. Simonton (Ed.), *The Wiley Handbook of Genius* (pp. 62–86). Oxford: Wiley.

Feldman, D. H. (1984). A follow-up of scoring above 180 IQ in Terman's "Genetic Studies of Genius." *Exceptional Children, 50,* 518–523.

Freeman, J. W. (1984). *The Metropolitan Opera: Stories of the Great Operas.* New York: W. W. Norton.

Finke, R. A., Ward, T. B., & Smith, S. M. (1992). *Creative Cognition: Theory, Research, Applications.* Cambridge, MA: MIT Press.

Galenson, D. W. (2001). *Painting Outside the Lines: Patterns of Creativity in Modern Art.* Cambridge, MA: Harvard University Press.

Galenson, D. W. (2005). *Old Masters and Young Geniuses: The Two Life Cycles of Artistic Creativity.* Princeton, NJ: Princeton University Press.

Galton, F. (1865). Hereditary talent and character. *Macmillan's Magazine, 12,* 157–166.

Galton, F. (1869). *Hereditary Genius: An Inquiry into its Laws and Consequences.* London: Macmillan.

Galton, F. (1874). *English Men of Science: Their Nature and Nurture.* London: Macmillan.

Galton, F. (1883). *Inquiries into Human Faculty and Its Development.* London: Macmillan.

Galton, F. (1889). *Natural Inheritance.* London: Macmillan.

Gelb, M. J. (2002). *Discover Your Genius: How to Think Like History's Ten Most Revolutionary Minds.* New York: HarperCollins.

Gingerich, O. (2005). *The Book Nobody Read: Chasing the Revolutions of Nicolaus Copernicus.* New York: Walker.

Ginsburgh, V., & Weyers, S. (2005). Creativity and life cycles of artists. *Journal of Cultural Economics, 30,* 91–107.

Ginsburgh, V., & Weyers, S. (2014). Evaluating excellence in the arts. In D. K. Simonton (Ed.), *The Wiley Handbook of Genius* (pp. 511–532). Oxford: Wiley.

Gladwell, M. (2008). *Outliers: The Story of Success.* New York: Little, Brown.

Godart, F., Maddux, W. W., Shipilov, A., & Galinsky, A. D. (2015). Fashion with a foreign flair: Professional experiences abroad facilitate the creative innovations of organizations. *Academy of Management Journal, 58,* 195–220.

Goertzel, M. G., Goertzel, V., & Goertzel, T. G. (1978). *300 Eminent Personalities: A Psychosocial Analysis of the Famous.* San Francisco: Jossey-Bass.

Götz, K. O., & Götz, K. (1979). Personality characteristics of successful artists. *Perceptual and Motor Skills, 49,* 919–924.

Grohman, M. G., Ivcevic, Z., Silvia, P., & Kaufman, S. B. (2017). The role of passion and persistence in creativity. *Psychology of Aesthetics, Creativity, and the Arts, 11,* 376–385.

Grosul, M., & Feist, G. J. (2014). The creative person in science. *Psychology of Aesthetics, Creativity, and the Arts, 8,* 30–43.

Guilford, J. P. (1967). *The Nature of Human Intelligence.* New York: McGraw-Hill.

Hadamard, J. (1945). *The Psychology of Invention in the Mathematical Field.* Princeton, NJ: Princeton University Press.

Hass, R. W., & Weisberg, R. W. (2009). Career development in two seminal American songwriters: A test of the equal odds rule. *Creativity Research Journal, 21,* 183–190.

Harris, J. R. (1995). Where is the child's environment: A group socialization theory of development. *Psychological Review, 102,* 458–489.

Hayes, J. R. (1989). *The Complete Problem Solver* (2nd ed.). Hillsdale, NJ: Erlbaum.

Hellmanzik, C. (2014). Prominent modern artists: Determinants of creativity. In D. K. Simonton (Ed.), *The Wiley Handbook of Genius* (pp. 564–585). Oxford: Wiley.

Helmreich, R. L., Spence, J. T., Beane, W. E., Lucker, G. W., & Matthews, K. A. (1980). Making it in academic psychology: Demographic and personality correlates of attainment. *Journal of Personality and Social Psychology, 39*, 896–908.

Helmreich, R. L., Spence, J. T., & Pred, R. S. (1988). Making it without losing it: Type A, achievement motivation, and scientific attainment revisited. *Personality and Social Psychology Bulletin, 14*, 495–504.

Helson, R. (1971). Women mathematicians and the creative personality. *Journal of Consulting and Clinical Psychology, 36*, 210–220.

Helson, R. (1980). The creative woman mathematician. In L. H. Fox, L. Brody, & D. Tobin (Eds.), *Women and the Mathematical Mystique* (pp. 23–54). Baltimore: Johns Hopkins University Press.

Helson, R. (1990). Creativity in women: Outer and inner views over time. In M. A. Runco & R. S. Albert (Eds.), *Theories of Creativity* (pp. 46–58). Newbury Park, CA: Sage.

Helson, R., & Crutchfield, R. S. (1970). Mathematicians: The creative researcher and the average Ph.D. *Journal of Consulting and Clinical Psychology, 34*, 250–257.

Hilts, V. L. (1975). *A Guide to Francis Galton's English Men of Science*. Philadelphia: American Philosophical Society.

Hollingworth, L. S. (1926). *Gifted Children: Their Nature and Nurture*. New York: Macmillan.

Hollingworth, L. S. (1942). *Children Beyond 180 IQ: Origin and Development*. Yonkers-on-Hudson, NY: World Book.

Horgan, J. (1996). *The End of Science*. New York: Addison Wesley.

Howe, M. J. A. (1999). *Genius Explained.* Cambridge: Cambridge University Press.

Isaacson, W. (2014). *The Innovators: How a Group of Hackers, Geniuses, and Geeks Created the Digital Revolution.* New York: Simon & Schuster.

Jamison, K. R. (1989). Mood disorders and patterns of creativity in British writers and artists. *Psychiatry, 52,* 125–134.

John, O. P. (1990). The "Big Five" factor taxonomy: Dimensions of personality in the natural language and questionnaires. In L. A. Pervin (Ed.), *Handbook of Personality: Theory and Research* (pp. 66–100). New York: Guilford Press.

Johnson, W., & Bouchard, T. J., Jr. (2014). Genetics of intellectual and personality traits associated with creative genius: Could geniuses be Cosmobian Dragon Kings? In D. K. Simonton (Ed.), *The Wiley Handbook of Genius* (pp. 269–296). Oxford: Wiley.

Jones, B. F., Reedy, E. J., & Weinberg, B. A. (2014). Age and scientific genius. In D. K. Simonton (Ed.), *The Wiley Handbook of Genius* (pp. 422–450). Oxford: Wiley.

Kaufman, J. C. (2003). The cost of the muse: Poets die young. *Death Studies, 27,* 813–822.

Kaufman, J. C. (Ed.). (2014). *Creativity and Mental Illness.* New York: Cambridge University Press.

Kaufman, J. C., & Sexton, J. D. (2006). Why doesn't the writing cure help poets? *Review of General Psychology, 10,* 268–282.

Kell, H. J., & Lubinski, D. (2014). The study of mathematically precocious youth at maturity: Insights into elements of genius. In D. K. Simonton (Ed.), *The Wiley Handbook of Genius* (pp. 397–421). Oxford: Wiley.

Kessler, R. C., Chiu, W. T., Demler, O., Merikangas, K. R., & Walters, E. E. (2005). Prevalence, severity, and comorbidity of 12-month DSM-IV disorders in the National Comorbidity Survey Replication. *Archives of General Psychiatry, 62,* 617–627.

Keyes, C. L. M., & Haidt, J. (Eds.). (2003). *Flourishing: Positive Psychology and the Life Well-Lived.* Washington, DC: American Psychological Association; 10.1037/10594-000.

Kinnier, R. T., Metha, A. T., Buki, L. P., & Rawa, P. M. (1994). Manifest value of eminent psychologists: A content analysis of their obituaries. *Current Psychology, 13,* 88–94.

Ko, Y., & Kim, J. (2008). Scientific geniuses' psychopathology as a moderator in the relation between creative contribution types and eminence. *Creativity Research Journal, 20,* 251–261.

Koestler, A. (1964). *The Act of Creation.* New York: Macmillan.

Kozbelt, A. (2005). Factors affecting aesthetic success and improvement in creativity: A case study of musical genres in Mozart. *Psychology of Music, 33,* 235–255.

Kozbelt, A. (2008). One-hit wonders in classical music: Evidence and (partial) explanations for an early career peak. *Creativity Research Journal, 20,* 179–195.

Kozbelt, A. (2014). Musical creativity over the lifespan. In D. K. Simonton (Ed.), *The Wiley Handbook of Genius* (pp. 451–472). Oxford: Wiley.

Kramer, P. D. (2006). *Freud: Inventor of the Modern Mind.* New York: Harper Perennial.

Kroeber, A. L. (1917). The superorganic. *American Anthropologist, 19,* 163–214.

Kroeber, A. L. (1944). *Configurations of culture growth.* Berkeley: University of California Press.

Kuhn, T. S. (1970). *The Structure of Scientific Revolutions* (2nd ed.). Chicago: University of Chicago Press.

Kyaga, S. (2015). *Creativity and Mental Illness: The Mad Genius in Question.* New York: Palgrave Macmillan.

Lamb, D., & Easton, S. M. (1984). *Multiple Discovery: The Pattern of Scientific Genius.* Avebury, UK: Avebury.

Langley, P., Simon, H. A., Bradshaw, G. L., & Zythow, J. M. (1987). *Scientific Discovery*. Cambridge, MA: MIT Press.

Lehman, H. C. (1953). *Age and Achievement*. Princeton, NJ: Princeton University Press.

Leung, A. K., Maddux, W. W., Galinsky, A. D., & Chiu, C. (2008). Multicultural experience enhances creativity: The when and how. *American Psychologist, 63*, 169–181.

Levin, S. G., & Stephan, P. E. (1999). Are the foreign born a source of strength for U.S. science? *Science, 285*, 1213–1214.

Lombroso, C. (1891). *The Man of Genius*. London: Scott.

Lotka, A. J. (1926). The frequency distribution of scientific productivity. *Journal of the Washington Academy of Sciences, 16*, 317–323.

Ludwig, A. M. (1995). *The Price of Greatness: Resolving the Creativity and Madness Controversy*. New York: Guilford Press.

Ludwig, A. M. (1998). Method and madness in the arts and sciences. *Creativity Research Journal, 11*, 93–101.

Lykken, D. T. (1998). The genetics of genius. In A. Steptoe (Ed.), *Genius and the Mind: Studies of Creativity and Temperament in the Historical Record* (pp. 15–37). New York: Oxford University Press.

Maddux, W. W., Adam, H., & Galinsky, A. D. (2010). When in Rome … learn why the Romans do what they do: How multicultural learning experiences facilitate creativity. *Personality and Social Psychology Bulletin, 36*, 731–741.

Martindale, C. (1990). *The Clockwork Muse: The Predictability of Artistic Styles*. New York: Basic Books.

Martindale, C. (1995). Fame more fickle than fortune: On the distribution of literary eminence. *Poetics, 23*, 219–234.

Matthews, K. A., Helmreich, R. L., Beane, W. E., & Lucker, G. W. (1980). Pattern A, achievement striving, and scientific merit: Does Pattern A help or hinder? *Journal of Personality and Social Psychology, 39*, 962–967.

McCann, S. J. H. (2001). The precocity-longevity hypothesis: Earlier peaks in career achievement predict shorter lives. *Personality and Social Psychology Bulletin, 27*, 1429–1439.

McCrae, R. R., & Greenberg, D. M. (2014). Openness to experience. In D. K. Simonton (Ed.), *The Wiley Handbook of Genius* (pp. 222–243). Oxford: Wiley.

McFarlan, D. (Ed.). (1989). *Guinness Book of World Records*. New York: Bantam.

McKay, A. S., & Kaufman, J. C. (2014). Literary geniuses: Their life, work, and death. In D. K. Simonton (Ed.), *The Wiley Handbook of Genius* (pp. 473–487). Oxford: Wiley.

Mednick, S. A. (1962). The associative basis of the creative process. *Psychological Review, 69*, 220–232.

Merton, R. K. (1961). Singletons and multiples in scientific discovery: A chapter in the sociology of science. *Proceedings of the American Philosophical Society, 105*, 470–486.

Moulin, L. (1955). The Nobel Prizes for the sciences from 1901–1950: An essay in sociological analysis. *British Journal of Sociology, 6*, 246–263.

Murray, C. (2003). *Human Accomplishment: The Pursuit of Excellence in the Arts and Sciences, 800 B.C. to 1950*. New York: HarperCollins.

Murray, C. (2014). Genius in world civilization. In D. K. Simonton (Ed.), *The Wiley Handbook of Genius* (pp. 486–608). Oxford: Wiley.

Murray, P. (Ed.). (1989). *Genius: The History of an Idea*. Oxford: Blackwell.

Naroll, R., Benjamin, E. C., Fohl, F. K., Fried, M. J., Hildreth, R. E., & Schaefer, J. M. (1971). Creativity: A cross-historical pilot survey. *Journal of Cross-Cultural Psychology, 2*, 181–188.

Ness, R. B. (2013). *Genius Unmasked*. New York: Oxford University Press.

Newell, A., & Simon, H. A. (1972). *Human Problem Solving*. Englewood Cliffs, NJ: Prentice-Hall.

Nickles, T. (2003). Evolutionary models of innovation and the Meno problem. In L. V. Shavinina (Ed.), *The International Handbook on Innovation* (pp. 54–78). New York: Elsevier Science.

Nisbett, R. E. (1968). Birth order and participation in dangerous sports. *Journal of Personality and Social Psychology, 8*, 351–353.

Ogburn, W. K., & Thomas, D. (1922). Are inventions inevitable? A note on social evolution. *Political Science Quarterly, 37*, 83–93.

Over, R. (1982). The durability of scientific reputation. *Journal of the History of the Behavioral Sciences, 18*, 53–61.

Patrick, S. (2013). *Awakening your Inner Genius*. Clearwater, FL: Oculus Publishers.

Plomin, R., & Daniels, D. (1987). Why are children in the same family so different from one another? *Behavioral and Brain Sciences, 10*, 1–60.

Plomin, R., DeFries, J. C., Knopik, V. S., & Neiderhiser, J. M. (2016). Top 10 replicated findings from behavioral genetics. *Perspectives on Psychological Science, 11*, 3–23.

Poincaré, H. (1921). *The Foundations of Science: Science and Hypothesis, the Value of Science, Science and Method* (G. B. Halstead, trans.). New York: Science Press.

Popper, K. (1963). *Conjectures and Refutations*. London: Routledge.

Post, F. (1994). Creativity and psychopathology: A study of 291 world-famous men. *British Journal of Psychiatry, 165*, 22–34.

Quetelet, A. (1968). *A Treatise on Man and the Development of His Faculties*. New York: Franklin. (Reprint of 1842 Edinburgh translation of 1835 French original)

Raskin, E. A. (1936). Comparison of scientific and literary ability: A biographical study of eminent scientists and men of letters of the nineteenth century. *Journal of Abnormal and Social Psychology, 31*, 20–35.

Robinson, A., & Simonton, D. K. (2014). Catharine Morris Cox Miles and the lives of others (1890–1984). In A. Robinson & J. L. Jolly (Eds.), *A Century of Contributions to Gifted Education: Illuminating Lives* (pp. 101–114). London: Routledge.

Robledo, I. C. (2016). *The Secret Principles of Genius: The Key to Unlocking Your Hidden Genius Potential*. Self-pub: Amazon Kindle Edition.

Rodgers, J. L., Cleveland, H. H., van den Oord, E., & Rowe, D. C. (2000). Resolving the debate over birth order, family size, and intelligence. *American Psychologist, 55*, 599–612.

Roe, A. (1952). A psychologist examines 64 eminent scientists. *Scientific American, 187*(5), 21–25.

Roe, A. (1953). *The Making of a Scientist*. New York: Dodd, Mead.

Rohrer, J. M., Egloff, B., & Schmukle, S. C. (2015). Examining the effects of birth order on personality. *Proceedings of the National Academy of Sciences of the United States of America, 112*, 14224–14229.

Root-Bernstein, R., Allen, L., Beach, L., Bhadula, R., Fast, J., Hosey, C., et al. (2008). Arts foster scientific success: Avocations of Nobel, National Academy, Royal Society, and Sigma Xi members. *Journal of the Psychology of Science and Technology, 1*, 51–63.

Rosengren, K. E. (1985). Time and literary fame. *Poetics, 14*, 157–172.

Rothenberg, A. (2015). *Flight from Wonder: An Investigation of Scientific Creativity*. Oxford: Oxford University Press.

Rushton, J. P. (1990). Creativity, intelligence, and psychoticism. *Personality and Individual Differences, 11*, 1291–1298.

Saad, C. S., Damian, R. I., Benet-Martinez, V., Moons, W. G., & Robins, R. W. (2013). Multiculturalism and creativity: Effects of cultural context, bicultural identity, and cognitive fluency. *Social Psychological & Personality Science, 4*, 369–375.

Schaefer, C. E., & Anastasi, A. (1968). A biographical inventory for identifying creativity in adolescent boys. *Journal of Applied Psychology, 58*, 42–48.

Schlesinger, J. (2009). Creative mythconceptions: A closer look at the evidence for the "mad genius" hypothesis. *Psychology of Aesthetics, Creativity, and the Arts, 3,* 62–72.

Schneider, J. (1937). The cultural situation as a condition for the achievement of fame. *American Sociological Review, 2,* 480–491.

Schubert, D. S. P., Wagner, M. E., & Schubert, H. J. P. (1977). Family constellation and creativity: Firstborn predominance among classical music composers. *Journal of Psychology, 95,* 147–149.

Simon, H. A. (1954). Spurious correlation: A causal interpretation. *Journal of the American Statistical Association, 49,* 467–479.

Simon, H. A. (1955). On a class of skew distribution functions. *Biometrika, 42,* 425–440.

Simon, H. A. (1956). Rational choice and the structure of the environment. *Psychological Review, 63,* 129–138.

Simon, H. A. (1973). Does scientific discovery have a logic? *Philosophy of Science, 40,* 471–480.

Simon, H. A. (1974). The structure of ill structured problems. *Artificial Intelligence, 4,* 181–201.

Simon, H. A., & Chase, W. G. (1973). Skill in chess. *American Scientist, 61,* 394–403.

Simonton, D. K. (1974). *The social psychology of creativity: An archival data analysis.* Unpublished doctoral dissertation, Harvard University.

Simonton, D. K. (1975a). Age and literary creativity: A cross-cultural and transhistorical survey. *Journal of Cross-Cultural Psychology, 6,* 259–277.

Simonton, D. K. (1975b). Sociocultural context of individual creativity: A transhistorical time-series analysis. *Journal of Personality and Social Psychology, 32,* 1119–1133.

Simonton, D. K. (1976a). Biographical determinants of achieved eminence: A multivariate approach to the Cox data. *Journal of Personality and Social Psychology, 33,* 218–226.

Simonton, D. K. (1976b). Philosophical eminence, beliefs, and zeitgeist: An individual-generational analysis. *Journal of Personality and Social Psychology, 34,* 630–640.

Simonton, D. K. (1977). Eminence, creativity, and geographic marginality: A recursive structural equation model. *Journal of Personality and Social Psychology, 35,* 805–816.

Simonton, D. K. (1979). Multiple discovery and invention: Zeitgeist, genius, or chance? *Journal of Personality and Social Psychology, 37,* 1603–1616.

Simonton, D. K. (1980). Techno-scientific activity and war: A yearly time-series analysis, 1500–1903 A.D. *Scientometrics, 2,* 251–255.

Simonton, D. K. (1984a). Artistic creativity and interpersonal relationships across and within generations. *Journal of Personality and Social Psychology, 46,* 1273–1286.

Simonton, D. K. (1984b). Leaders as eponyms: Individual and situational determinants of monarchal eminence. *Journal of Personality, 52,* 1–21.

Simonton, D. K. (1986). Biographical typicality, eminence, and achievement style. *Journal of Creative Behavior, 20,* 14–22.

Simonton, D. K. (1988a). Age and outstanding achievement: What do we know after a century of research? *Psychological Bulletin, 104,* 251–267.

Simonton, D. K. (1988b). Galtonian genius, Kroeberian configurations, and emulation: A generational time-series analysis of Chinese civilization. *Journal of Personality and Social Psychology, 55,* 230–238.

Simonton, D. K. (1989). Shakespeare's sonnets: A case of and for single-case historiometry. *Journal of Personality, 57,* 695–721.

Simonton, D. K. (1990). Lexical choices and aesthetic success: A computer content analysis of 154 Shakespeare sonnets. *Computers and the Humanities, 24,* 251–264.

Simonton, D. K. (1991a). Career landmarks in science: Individual differences and interdisciplinary contrasts. *Developmental Psychology, 27,* 119–130.

Simonton, D. K. (1991b). Emergence and realization of genius: The lives and works of 120 classical composers. *Journal of Personality and Social Psychology, 61*, 829–840.

Simonton, D. K. (1991c). Latent-variable models of posthumous reputation: A quest for Galton's G. *Journal of Personality and Social Psychology, 60*, 607–619.

Simonton, D. K. (1992a). Leaders of American psychology, 1879–1967: Career development, creative output, and professional achievement. *Journal of Personality and Social Psychology, 62*, 5–17.

Simonton, D. K. (1992b). The social context of career success and course for 2,026 scientists and inventors. *Personality and Social Psychology Bulletin, 18*, 452–463.

Simonton, D. K. (1994). *Greatness: Who Makes History and Why*. New York: Guilford Press.

Simonton, D. K. (1997a). Creative productivity: A predictive and explanatory model of career trajectories and landmarks. *Psychological Review, 104*, 66–89.

Simonton, D. K. (1997b). Foreign influence and national achievement: The impact of open milieus on Japanese civilization. *Journal of Personality and Social Psychology, 72*, 86–94.

Simonton, D. K. (1998). Fickle fashion versus immortal fame: Transhistorical assessments of creative products in the opera house. *Journal of Personality and Social Psychology, 75*, 198–210.

Simonton, D. K. (2000a). Creative development as acquired expertise: Theoretical issues and an empirical test. *Developmental Review, 20*, 283–318.

Simonton, D. K. (2000b). Creativity: Cognitive, developmental, personal, and social aspects. *American Psychologist, 55*, 151–158.

Simonton, D. K. (2000c). Methodological and theoretical orientation and the long-term disciplinary impact of 54 eminent psychologists. *Review of General Psychology, 4*, 13–24.

Simonton, D. K. (2001). Predicting presidential greatness: Equation replication on recent survey results. *Journal of Social Psychology, 141,* 293–307.

Simonton, D. K. (2004). *Creativity in Science: Chance, Logic, Genius, and Zeitgeist.* Cambridge: Cambridge University Press.

Simonton, D. K. (2006). Presidential IQ, Openness, Intellectual Brilliance, and leadership: Estimates and correlations for 42 US chief executives. *Political Psychology, 27,* 511–526.

Simonton, D. K. (2007a). Creative life cycles in literature: Poets versus novelists or conceptualists versus experimentalists? *Psychology of Aesthetics, Creativity, and the Arts, 1,* 133–139.

Simonton, D. K. (2007b). The creative process in Picasso's *Guernica* sketches: Monotonic improvements or nonmonotonic variants? *Creativity Research Journal, 19,* 329–344.

Simonton, D. K. (2008a). Childhood giftedness and adulthood genius: A historiometric analysis of 291 eminent African Americans. *Gifted Child Quarterly, 52,* 243–255.

Simonton, D. K. (2008b). Gender differences in birth order and family size among 186 eminent psychologists. *Journal of Psychology of Science and Technology, 1,* 15–22.

Simonton, D. K. (2008c). Scientific talent, training, and performance: Intellect, personality, and genetic endowment. *Review of General Psychology, 12,* 28–46.

Simonton, D. K. (2009). The "other IQ": Historiometric assessments of intelligence and related constructs. *Review of General Psychology, 13,* 315–326.

Simonton, D. K. (2010). Creativity as blind-variation and selective-retention: Combinatorial models of exceptional creativity. *Physics of Life Reviews, 7,* 156–179.

Simonton, D. K. (2011). *Great Flicks: Scientific Studies of Cinematic Creativity and Aesthetics.* New York: Oxford University Press.

Simonton, D. K. (2012a). Foresight, insight, oversight, and hindsight in scientific discovery: How sighted were Galileo's telescopic sightings? *Psychology of Aesthetics, Creativity, and the Arts, 6*, 243–254.

Simonton, D. K. (2012b). The science of genius. *Scientific American Mind, 23*(5), 34–41.

Simonton, D. K. (2013a). After Einstein: Scientific genius is extinct. *Nature, 493*, 602.

Simonton, D. K. (2014a). Creative performance, expertise acquisition, individual-differences, and developmental antecedents: An integrative research agenda. *Intelligence, 45*, 66–73.

Simonton, D. K. (2014b). The mad-genius paradox: Can creative people be more mentally healthy but highly creative people more mentally ill? *Perspectives on Psychological Science, 9*, 470–480.

Simonton, D. K. (2014c). More method in the mad-genius controversy: A historiometric study of 204 historic creators. *Psychology of Aesthetics, Creativity, and the Arts, 8*, 53–61.

Simonton, D. K. (Ed.). (2014d). *The Wiley Handbook of Genius*. Oxford: Wiley.

Simonton, D. K. (2015a). Psychology as a science within Comte's hypothesized hierarchy: Empirical investigations and conceptual implications. *Review of General Psychology, 19*, 334–344.

Simonton, D. K. (2015b). Thomas Alva Edison's creative career: The multilayered trajectory of trials, errors, failures, and triumphs. *Psychology of Aesthetics, Creativity, and the Arts, 9*, 2–14.

Simonton, D. K. (2016a). Creativity, automaticity, irrationality, fortuity, fantasy, and other contingencies: An eightfold response typology. *Review of General Psychology, 20*, 194–204.

Simonton, D. K. (2016b). The decline of the West? A comparative civilizations perspective. In D. Ambrose & R. J. Sternberg (Eds.), *Creative Intelligence in the 21st Century: Grappling with Enormous Problems and Huge Opportunities* (pp. 51–64). Rotterdam, Netherlands: Sense Publishers.

Simonton, D. K. (2016c). Early and late bloomers among classical composers: Were the greatest geniuses also prodigies? In G. McPherson (Ed.), *Musical prodigies: Interpretations from psychology, music education, musicology and ethnomusicology* (pp. 185–197). New York: Oxford University Press.

Simonton, D. K. (2016d). Scientific genius in Islamic civilization: Quantified time series from qualitative historical narratives. *Journal of Genius and Eminence, 1*, 4–13.

Simonton, D. K. (2017a). Eminent female psychologists in family context: Historical trends for 80 women born 1847–1950. *Journal of Genius and Eminence, 1*(2), 15–25.

Simonton, D. K. (2017b). Intellectual genius in the Islamic Golden Age: Cross-civilization replications, extensions, and modifications. *Psychology of Aesthetics, Creativity, and the Arts*; Advance online publication. doi:10.1037/aca0000110.

Simonton, D. K., & Damian, R. I. (2013). Creativity. In D. Reisberg (Ed.), *Oxford Handbook of Cognitive Psychology* (pp. 795–807). New York: Oxford University Press.

Simonton, D. K., & Song, A. V. (2009). Eminence, IQ, physical and mental health, and achievement domain: Cox's 282 geniuses revisited. *Psychological Science, 20*, 429–434.

Simonton, D. K., & Ting, S.-S. (2010). Creativity in Eastern and Western civilizations: The lessons of historiometry. *Management and Organization Review, 6*, 329–350.

Sinatra, R., Wang, D., Deville, P., Song, C., & Barabási, A.-L. (2016). Quantifying the evolution of individual scientific impact. *Science, 354*. doi:10.1126/science.aaf5239.

Skinner, B. F. (1981). Selection by consequences. *Science, 213*, 501–504.

Song, A. V., & Simonton, D. K. (2007). Personality assessment at a distance: Quantitative methods. In R. W. Robins, R. C. Fraley, & R. F. Krueger (Eds.), *Handbook of Research Methods in Personality Psychology* (pp. 308–321). New York: Guilford Press.

Sorokin, P. A. (1937). *Social and Cultural Dynamics* (Vols. *1–4*). New York: American Book.

Sorokin, P. A. (1963). *A Long Journey: The Autobiography of Pitirim A. Sorokin*. New Haven, CT: College and University Press.

Sternberg, R. J., & Bridges, S. L. (2014). Varieties of genius. In D. K. Simonton (Ed.), *The Wiley Handbook of Genius* (pp. 185–197). Oxford: Wiley.

Stewart, L. H. (1977). Birth order and political leadership. In M. G. Hermann (Ed.), *The Psychological Examination of Political Leaders* (pp. 205–236). New York: Free Press.

Suedfeld, P. (2014). Political and military geniuses: Psychological profiles and responses to stress. In D. K. Simonton (Ed.), *The Wiley Handbook of Genius* (pp. 244–266). Oxford: Wiley.

Sulloway, F. J. (1996). *Born to Rebel: Birth Order, Family Dynamics, and Creative Lives*. New York: Pantheon.

Sulloway, F. J. (2007). Birth order and intelligence. *Science, 317,* 1711–1712.

Sulloway, F. J. (2010). Why siblings are like Darwin's Finches: Birth order, sibling competition, and adaptive divergence within the family. In D. M. Buss & P. H. Hawley (Eds.), *The Evolution of Personality and Individual Differences* (pp. 86–119). Oxford: Oxford University Press.

Sulloway, F. J. (2014). Openness to scientific innovation. In D. K. Simonton (Ed.), *The Wiley Handbook of Genius* (pp. 546–563). Oxford: Wiley.

Sulloway, F. J., & Zweigenhaft, R. L. (2010). Birth order and risk taking in athletics: A meta-analysis and study of major league baseball players. *Personality and Social Psychology Review, 14*, 402–416.

Teigen, K. H. (1984). A note on the origin of the term "nature and nurture": Not Shakespeare and Galton, but Mulcaster. *Journal of the History of the Behavioral Sciences, 20*, 363–364.

Terman, L. M. (1916). *The Measurement of Intelligence: An Explanation of and a Complete Guide for the Use of the Stanford Revision and Extension of the Binet-Simon Intelligence Scale*. Boston: Houghton Mifflin.

Terman, L. M. (1917). The intelligence quotient of Francis Galton in childhood. *American Journal of Psychology, 28*, 209–215.

Terman, L. M. (1925–1959). *Genetic Studies of Genius* (5 vols.). Stanford, CA: Stanford University Press.

Terman, L. M. (1925). *Mental and Physical Traits of a Thousand Gifted Children*. Stanford, CA: Stanford University Press.

Terman, L. M., & Oden, M. H. (1947). *The Gifted Child Grows Up*. Stanford, CA: Stanford University Press.

Terman, L. M., & Oden, M. H. (1959). *The Gifted Group at Mid-life*. Stanford, CA: Stanford University Press.

Terry, W. S. (1989). Birth order and prominence in the history of psychology. *Psychological Record, 39*, 333–337.

Thagard, P. (2012). Creative combination of representations: Scientific discovery and technological invention. In R. Proctor & E. J. Capaldi (Eds.), *Psychology of Science: Implicit and Explicit Processes* (pp. 389–405). New York: Oxford University Press.

Tomlinson-Keasey, C. (1990). The working lives of Terman's gifted women. In H. Y. Grossman & N. L. Chester (Eds.), *The Experience and Meaning of Work in Women's Lives* (pp. 213–239). Hillsdale, NJ: Erlbaum.

Turkheimer, E., & Waldron, M. (2000). Nonshared environment: A theoretical, methodological, and quantitative review. *Psychological Bulletin, 126*, 78–108.

Walters, J., & Gardner, H. (1986). The crystallizing experience: Discovering an intellectual gift. In R. J. Sternberg & J. E. Davidson (Eds.), *Conceptions of Giftedness* (pp. 306–331). New York: Cambridge University Press.

Weisberg, R. W. (2004). On structure in the creative process: A quantitative case-study of the creation of Picasso's *Guernica*. *Empirical Studies of the Arts, 22*, 23–54.

Weisberg, R. W. (2014). Case studies of genius: Ordinary thinking, extraordinary outcomes. In D. K. Simonton (Ed.), *The Wiley Handbook of Genius* (pp. 139–165). Oxford: Wiley.

West, S. S. (1960). Sibling configurations of scientists. *American Journal of Sociology, 66*, 268–274.

White, R. K. (1931). The versatility of genius. *Journal of Social Psychology, 2*, 460–489.

Whitehead, A. N. (1978). D. R. Griffin & D. W. Sherburne (Eds.), *Process and Reality: An Essay in Cosmology*. New York: Free Press. (Original work published 1929)

Wiener, N. (1953). *Ex-prodigy: My Childhood and Youth*. New York: Simon & Schuster.

Winner, E. (2014). Child prodigies and adult genius: A weak link. In D. K. Simonton (Ed.), *The Wiley Handbook of Genius* (pp. 297–320). Oxford: Wiley.

Wirth, E. (2008). *Thomas Edison in West Orange*. Charleston, SC: Arcadia Publishing.

Wohl, C. G. (2007). Scientist as detective: Luis Alvarez and the pyramid burial chambers, the JFK assassination, and the end of the dinosaurs. *American Journal of Physics, 75*, 968–977. doi:10.1119/1.2772290.

Zajonc, R. B. (1976). Family configuration and intelligence. *Science, 192*, 227–235.

Zajonc, R. B. (1983). Validating the confluence model. *Psychological Bulletin, 93*, 457–480.

Zuckerman, H. (1977). *Scientific Elite*. New York: Free Press.

Zusne, L. (1976). Age and achievement in psychology: The harmonic mean as a model. *American Psychologist, 31*, 805–807.

Index